ICE Specification for piling and embedded retaining walls
2nd edition

Institution of Civil Engineers

Association of Geotechnical & Geoenvironmental Specialists

· THE BRITISH ·
GEOTECHNICAL
ASSOCIATION

THE GROUND FORUM

HIGHWAYS
AGENCY

thomas telford

Published by Thomas Telford Publishing, Thomas Telford Ltd, 1 Heron Quay, London E14 4JD. http://www.thomastelford.com

Distributors for Thomas Telford books are
USA: ASCE Press, 1801 Alexander Bell Drive, Reston, VA 20191-4400, USA
Japan: Maruzen Co. Ltd, Book Department, 3–10 Nihonbashi 2-chome, Chuo-ku, Tokyo 103
Australia: DA Books and Journals, 648 Whitehorse Road, Mitcham 3132, Victoria

First published 2007
Reprinted with amendment 2007

Also available from Thomas Telford Books

The Essential Guide to the ICE Specification for Piling and Embedded Retaining Walls. ISBN 07277 2738 9. The Federation of Piling Specialists

The Specification for Piling and Embedded Retaining Walls. ISBN 07277 2566 1. Institution of Civil Engineers

A catalogue record for this book is available from the British Library

ISBN: 978-0-7277-3358-0

Typeset by Academic + Technical, Bristol
Printed and bound in Great Britain by MPG Books, Bodmin, Cornwall

Foreword

Welcome to the latest edition of the ICE Specification for Piling and Embedded Retaining Walls – the third in a highly successful series, some eleven years after the first edition of the complete ICE Specification for Piling and Embedded Retaining Walls (SPERW) and almost twenty years after the original ICE Specification for Piling. This specification has been one of the most popular of the ICE's standard specifications and has been influential in raising piling standards in the UK and in other places where it has been used. The vast majority of UK piling contracts now refer to it (or to derivatives via the National Building Specification or the Highways Agency's Specification for Highway Works), and its requirements are well known to most specialist consultants and contractors. The existence of a standard agreed way of executing piling works helps to reduce conflict on site and enables consultants to design more economically by having confidence in how site works will be carried out.

For those interested in the history of the evolution of the document, it is based on the following documents:

- ICE Specification for Piling (1988)
- Draft Specification for Embedded Retaining Walls – unpublished
- Highways Agency Specification for Piling and Embedded Retaining Walls (1994)
- ICE Specification for Piling and Embedded Retaining Walls (1996)
- FPS Essential Guide to the ICE Specification for Piling and Embedded Retaining Walls (1999).

The document has a new look to make it more usable. Part A is an introduction to the essential concepts necessary to procure a piling contract. Part B is the specification and is the only part of this document intended for incorporation in contracts. Part C provides guidance for use of the specification and essential background information for specifiers and contractors alike. This is an advance on the previous first edition of SPERW where the guidance was dispersed. This new edition also incorporates the helpful commentary produced by the Federation of Piling Specialists ('the essential guide' from 1999) so contained in one authoritative document are all the guidance documents needed to prepare and work to this piling specification.

In terms of national standards, much has happened since the publication of the previous edition. The Eurocodes have been written and are now in widespread usage. Many European standards have now been issued as British Standards. This new edition has striven to keep pace with these fast-moving changes. Likewise, safety standards continue to improve and the piling industry has responded positively to each new initiative.

This specification has been written by practising engineers for practising engineers. It is the product of wide consultation and rigorous review to ensure that it takes account of latest developments and updates to standards. The review panel has been drawn from consultants, specialist foundation contractors and client organisations.

Comments have been widely sought and the penultimate draft was reviewed by specialists to ensure consistency and clarity. Accordingly, this edition represents a major advance and should facilitate even more rapid development of the piling and embedded walling industry. It is therefore commended to the UK construction industry.

Finally, the ICE would like to gratefully acknowledge those individuals who have contributed so much to the development of the specification. Consultants and contractors have collaborated to produce a compromise between a desire for exacting standards and an economical approach that balances risk and cost. It is this long collaboration between many highly-skilled people that makes this specification so valuable. I would like to add my thanks to the contributors for this excellent new edition of the specification.

Professor Michael C. R. Davies
Chairman, British Geotechnical Association

Acknowledgements

ICE *Specification for piling*, **1988**
J. Bickerdike; A. Fawcett; K. Fleming; J. May; I. McFarlane;
D. Palmer; S. Thorburn; D. Wake; J. Woodhouse.

ICE Working Group for draft *Specification for embedded retaining walls*
F. Chartres; J. Findlay (Chairman); E. Haws; D. Sherwood; V. Troughton.

Highways Agency *Specification for piling and embedded retaining walls*, **1994**
D. Bush; T. Chapman; K. Cole; B. Simpson; V. Troughton (Chairman);
D. Twine. This specification benefited from earlier contributions by
J. Mitchell and A. Turner.

FPS Mirror Group for Highways Agency *Specification*
A. Fawcett; J. Findlay; K. Fleming (Chairman); L. Stansfield.

ICE *Specification for piling and embedded retaining walls*, **1996**
T. Chapman (Chairman); J. Findlay; A. Lord; M. Alexander.

FPS *The essential guide to the ICE Specification for piling and embedded retaining walls*, **1999**
D. Evans; K. Fleming; D. Illingworth; P. McIvor; N. Mure; Q. Spear;
L. Stansfield; V. Troughton (Chairman); S. Wade; C. Whalley.

ICE *Specification for piling and embedded retaining walls*, **2007**
R. Fernie (Chairman), FPS/Ground Forum; E. Evans (Steering group/principal authors), Network Rail; A. Kidd (Steering group/principal authors), Highways Agency; S. Lee (Steering group/principal authors), Network Rail; T. Chapman (Steering group/principal authors), Arup; J. Cook (Steering group/principal authors), AGS/Ground Forum; P. Ingram (Steering group/principal authors), Arup; D. Puller (Steering group/principal authors), FPS/BGA; T. Suckling (Steering group/principal authors), FPS/BGA; O. Synac (Steering group/principal authors), FPS; D. Corke (Technical Reviewer), DC Project Solutions; T. Butcher (Specialist contributors to individual chapters), BRE; A. Lawrence (Specialist contributors to individual chapters), Arup; B. Marsh (Specialist contributors to individual chapters), Arup; P. Ross (Specialist contributors to individual chapters), Arup; D. Rowbottom (Specialist contributors to individual chapters), Steel Piling Group; G. White (Specialist contributors to individual chapters), Steel Piling Group.

These are the 'headline' authors/reviewers. The steering group acknowledges the interest and effort expanded by many other members of the participating teams and the interest and help afforded generally by practicing colleagues and contemporaries.

Financial support from Network Rail, the Highways Agency and Thomas Telford is gratefully acknowledged. The authors and their organizations have also supported the production of this document by generous donation of their time.

Contents

Introduction

The Institution of Civil Engineers (ICE) *Specification for piling and embedded retaining walls* 2007 (*SPERW*) is the latest version of this successful document for piling and embedded walling works. Earlier versions were published in 1988 and in 1996.

SPERW has been updated to reflect the latest piling techniques and procurement methods used in the United Kingdom (UK) foundations market, and the introduction of European Standards. *SPERW* is intended for use as a technical specification for piling and embedded walling works either on land or near to shore. It is not applicable for offshore works. Piles or shafts constructed using hand-dug methods are specifically excluded from *SPERW*.

SPERW comprises three parts as follows:

Part A – General requirements
Part B – Specification requirements
Part C – Guidance notes

Part A is general guidance describing the requirements typically necessary for the successful construction of piling and embedded retaining walls. Information is given on the tendering process, design and on issues such as safety, quality and the environment.

Part B is the main technical Specification advocated for use on piling and embedded walling works in the UK. Part B comprises 19 sections covering the main piling and embedded walling methods, and the most common testing methods and materials used in these types of foundations works. Definitions are given in Section B1.18.

Part C provides specific guidance on the use of each of the 19 sections within the Specification Part B. This guidance collates the experiences of contractors and consultants involved for many years with piling and embedded retaining walls.

It is intended that this document continues to be used as the UK national standard Specification for piling and embedded retaining walls and that specifiers will continue to refer only to *SPERW* so that its standard clauses need not be reproduced for every contract. Specifiers can make special amendments to *SPERW* with the use of a Project Specification containing the details described within each of the sections in Part B.

Specifiers should refer only to the relevant sections within Part B that are applicable for the foundation type, as indicated by the asterixes in Table 1. In this document the walling Sections 8 to 12 should be specified in conjunction with the appropriate piling method sections; for example to specify a contiguous concrete pile wall Sections B1, B3 and/or B4, B10 and B19 should be referred to.

Parts A and C of *SPERW* are not intended to form part of any contractual Specification, but the guidance contained in Parts A and C would normally be reflected within the contract documentation. It is intended that only Part B of *SPERW* should form part of the contractual Specification.

Part B of *SPERW* is the technical Specification and is intended to be used with any form of contract. Accordingly it does not contain detailed guidance on contract documentation and measurement.

Table 1 Sections in Part B that may be relevant for bearing pile works or for embedded retaining wall works

Section no. in Part B	1	2	3	4	5	6	7	8	9	10	11	12	13	14	15	16	17	18	19
Title	Specification requirements	Driven pre-cast concrete piles	Bored concrete piles	CFA and DA concrete piles	Driven concrete piles	Steel piles	Timber piles	Diaphragm walls and barrettes	Secant pile walls	Contiguous pile walls	King post pile walls	Sheetpile walls	Integrity testing of piles	Dynamic testing of piles	Static load testing of piles	Friction reduction	Instrumentation	Support fluid	Concrete and reinforcement
Bearing piles	*	*	*	*	*	*	*	*					*	*	*	*	*	*	*
Embedded retaining walls	*	*	*	*	*	*	*	*	*	*	*	*	*				*	*	*

This is because of the plethora of different contract forms now regularly in use in the UK foundations market. For further guidance on contract documentation and measurement relevant for piling and embedded retaining walls, see www.fps.org.uk.

Part B of *SPERW* relates to materials and workmanship and is not a design document. Every effort has been made to avoid conflict between this Specification, the ICE conditions, the JCT conditions and other forms of contract. However, certain clauses in the ICE conditions do not have parallel clauses in other conditions. Therefore, clauses in this Specification, which the Engineer considers are covered by the contract conditions, should be amended or deleted.

There are many European Standards currently being introduced which will work in parallel with BS EN 1997 *Geotechnical design*; see Table 2 for some Standards relevant for piling and embedded

Table 2 Some European Standards relevant for piling and embedded retaining wall works

Type of European Standard	EN no.	Title
Eurocode	1992 Part 1-1	Design of Concrete Structures: General Rules and Rules for Buildings
Eurocode	1993 Part 5	Design of Steel Structures – Piling
Eurocode	1995 Part 1-1	Design of Timber Structures: General – Common Rules and Rules for Buildings
Eurocode	1997 Part 1	Geotechnical Design – General Rules
Eurocode	1997 Part 2	Geotechnical Design – Ground Investigation and Testing
Execution	1536	Execution of Special Geotechnical Work – Bored Piles
Execution	1538	Execution of Special Geotechnical Work – Diaphragm Walls
Execution	12063	Execution of Special Geotechnical Work – Sheet Piles
Execution	12699	Execution of Special Geotechnical Work – Displacement Piles
Execution	12715	Execution of Special Geotechnical Work – Grouting
Execution	14199	Execution of Special Geotechnical Work – Micropiles
Testing	22477 Part 1	Testing of Piles – Static Axially Loaded Compression Test
Testing	22477 Part 2	Testing of Piles – Static Axially Loaded Tension Test
Testing	22477 Part 3	Testing of Piles – Static Transversally Loaded Tension Test
Testing	22477 Part 4	Testing of Piles – Dynamic Axially Loaded Compression Test
Harmonized British Standard	12794	Precast Concrete Products – Foundation Piles

retaining walls. Some of these have been published and some are still being prepared.

The Eurocode Standards provide common structural design rules for everyday use in the design of whole structures and component products of both a traditional and an innovative nature. The Eurocode programme comprises ten Standards which generally consist of a number of parts. In addition to the Eurocodes, there are several European Standards that cover the execution procedures for geotechnical works which have been prepared to stand alongside BS EN 1997-1 as well as a range of CEN Standards which address testing procedures. Also some British Standards have been harmonized with the European Standards so that they now comply with them.

Monitoring of the development of all Standards independent of *SPERW* should be undertaken by the Engineer before writing any Project Specification.

This document has been designed for use with common practices, but is not intended to inhibit innovation. Novel solutions, such as the use of ribbed piles, piles with shaft enlargements or enlarged heads to piles, can be used with this Specification provided that additional clauses are included in the Project Specification which will ensure that the final product is constructed in accordance with the design requirements.

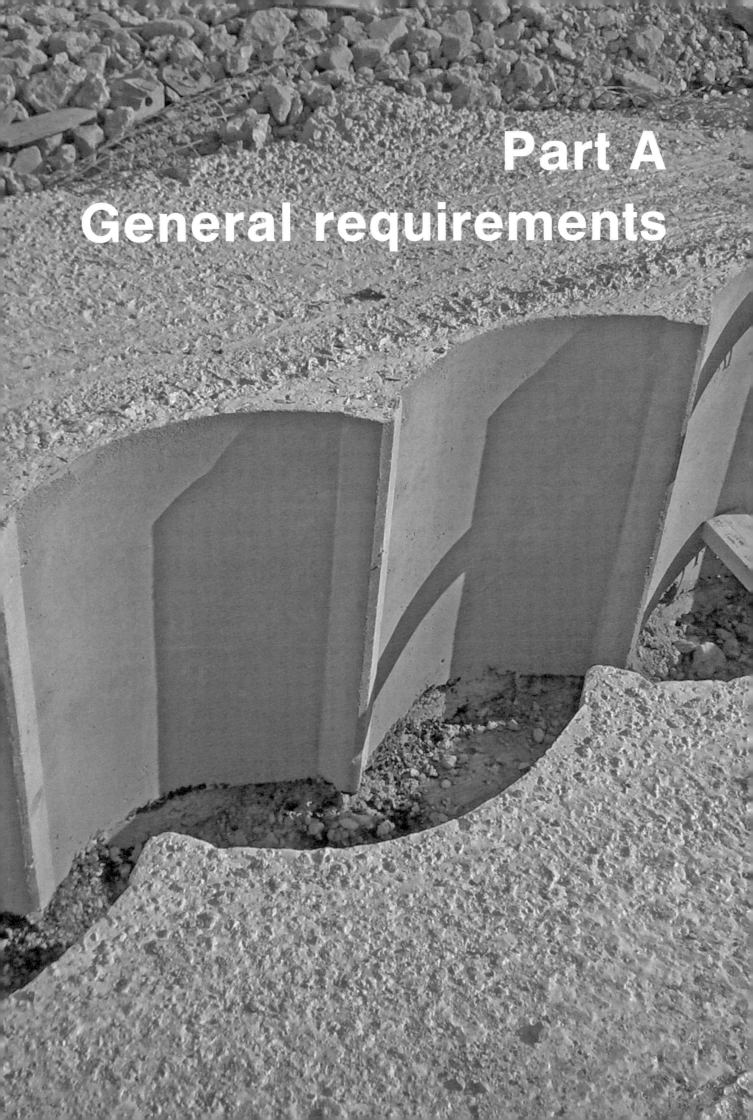

Part A
General requirements

General requirements

A1 Introduction

This part of the document is general guidance describing the requirements typically necessary for successful piling or embedded retaining wall works and is not intended to form part of any contractual specification. Information is given on the tendering process and on issues such as safety, quality and the environment.

A2 Specification roles

Within the Specification reference is made to the roles of the 'Engineer' and the 'Contractor'. As this Specification is published by the ICE, the 'Engineer' is assumed to have full delegated powers to control the contract and is acting on behalf of the 'Employer'. The two parties to the contract are the 'Employer' and the 'Contractor'.

Where piling or embedded walling works are executed under a contract which is not governed by ICE conditions, the Project Specification should state which bodies are to be nominated in their place. Where the Contractor has design responsibility under the terms of the contract, the Project Specification should state the role of the Engineer. An example Project Specification is given in Appendix 1.

The term 'Contractor' is always deemed to mean the principal or main contractor appointed by the Employer to undertake the contract works. This could be a specialist piling or walling contractor where they are appointed directly by the Employer. This document specifies only the direct contractual responsibilities between the Employer and their Contractor.

It is recognized that under different contract philosophies the 'Contractor' may opt to delegate some or all of their responsibilities to other organizations such as subcontractors. Responsibilities between the Contractor and their subcontractors and suppliers should be specified in the subcontract documentation and, in the absence of any provisions to the contrary, the 'Contractor' will always be taken to mean the party that is in contract with the Employer.

In Part C of this document the role of 'Designer' is introduced alongside the 'Engineer' and the 'Contractor'. This is to help clarify responsibilities as, for the design of piling or embedded walling works, either the 'Engineer' or the 'Contractor' can be responsible for the design of all, or part of, the foundation scheme, the choice of method, and the piles or wall elements.

A3 Responsibilities for safety, quality and the environment
A3.1 Safety
A3.1.1 Design and construction

Piling is potentially a dangerous activity and, as a minimum, the Contractor must carry out the works in compliance with the requirements of UK law, and pay particular regard to the Health and Safety at Work Act and other safety legislation that is in force and is applicable to the health and safety of persons involved with or affected by the execution of the contract.

The design and construction of the works shall be carried out in accordance with the latest safety legislation including the Construction (Design & Management) Regulations (CDM). All construction plant must be maintained and operated in a safe manner.

A3.1.2 Heath and Safety Plan

The Contractor should submit a Health and Safety Plan with the tender, which should include the following:

(*a*) The Contractor's Health and Safety Policy Statement.

(*b*) A general indication of the number and types of employees intended to undertake the contract.

(*c*) Brief details of the experience and relevant technical qualifications of the Contractor's Manager/Supervisor responsible for undertaking the contract.

(*d*) A method statement detailing how the work will be carried out safely. This shall take account of any hazard information supplied by the Employer.

(*e*) An assessment of any safety systems and general induction training that would be required before work can start on site and how this will be provided.

(*f*) Accident reporting arrangements.

(*g*) A summary of details of any major injuries and accidents reportable as required by The Reporting of Injuries, Diseases and Dangerous Occurrences Regulations 1995 that have occurred to the Contractor's personnel within the past two years.

(*h*) A broad description of the types of work carried out in the past two years by the Contractor.

(*i*) Additional requirements specifically mentioned in the invitation to tender.

Immediately prior to starting work for the first time the Contractor should make any necessary amendments to the tender Health and Safety Plan which should be agreed with the Engineer. After starting work on site the Contractor should formally review the Health and Safety Plan at suitable intervals to ensure the Health and Safety Plan is kept up to date and relevant to the work in hand.

It is the responsibility of the Contractor to bring all the relevant safety clauses to the attention of all the Contractor's personnel, including subcontractors and suppliers, as applicable to their work, and to ensure compliance with the contract safety clauses.

It is also the responsibility of the Contractor to ensure that any subcontractor or supplier employed by them on the project ensures that their personnel are aware of, and comply with, all relevant clauses. In particular, copies of relevant extracts of the Health and Safety Plan and method statements should be made readily available at the site for display by the subcontractor to their personnel.

A3.1.3 Risk management

Risk management is the identification, measurement and economic control of risks. A qualitative assessment of risk is essential in order to gain an appreciation of the relative importance of the various issues and to determine a risk management strategy.

Both the Engineer and the Contractor should undertake risk assessments for the piling or embedded walling works. The risk assessment process needs to be practical and take account of the views of their own staff and, where applicable, subcontractors and suppliers, who will all have practical knowledge to contribute.

For piling and embedded walling works it is essential that the risk management process continues after tendering throughout the construction process until the works are complete.

A3.1.4 Working platforms

Working platforms should be designed, constructed, maintained and repaired in accordance with BRE Report BR470. Further guidance on the use of hydraulically bound material (HBM) for working platforms has also been produced by BRE.

The Contractor should provide the platform designer with plant bearing pressures for all loading cases. The design of the working platform should take account of these plant bearing pressures and the ground conditions upon which the platform will be constructed. This design must be undertaken by a competent person with appropriate geotechnical expertise.

A Working Platform Certificate (see www.fps.org.uk), or similar, should be used to identify responsibilities and confirm that the working platform has been properly designed, constructed in accordance with this design, and will be adequately maintained and repaired to ensure the integrity of the platform throughout its working life.

It is recognized that preliminary working platform design is often undertaken for cost and design purposes prior to the appointment of the Contractor. In this instance all assumptions must be verified and, if necessary, the design revised prior to the construction of the working platform and the issue of the Working Platform Certificate.

A3.1.5 Pile trimming

Traditionally, concrete piles have been broken down by hand operated tools to limit the damage induced in the pile. In recent years the hazards of vibration-induced conditions such as hand-arm and whole-body vibration syndrome caused by these tools have come to the fore as a health issue for the construction industry. The Control of Vibration at Work Regulations are now in place and apply in combination with the CDM regulations. Hence both the Engineer and Contractor must comply with these requirements to reduce the risk to workers of vibration-induced conditions.

The Engineer is encouraged to design the structure so that pile trimming is minimized. The Engineer should then facilitate the use of improved technologies to ensure that risk from the exposure of site workers to vibration is either eliminated at source or, where this is not reasonably practicable, is reduced to a level which is as low as is reasonably practicable.

There are several methods that can be specified to help reduce this risk (see www.fps.org.uk). These systems may require the reinforcement above cut-off level to be debonded from the concrete. The applicability of these methods and any impact on tolerances should be agreed with the Contractor at tender stage. Alternatively, consideration should be given to the use of steel piling methods. Whilst trimming of steel piles will not provide a vibration induced risk to operatives, there will be other safety risks which will need to be taken into account.

A3.2 Quality
A3.2.1 Essential design requirements

Foundation design needs to fulfil three essential criteria in all cases:

- To have an adequate safety margin against failure.
- To have acceptable displacements over the likely range of applied loads.
- To be durable for the stated design life.

A3.2.2 Design and construction

It is essential that the method of construction is compatible with the design. This applies to both practicality aspects as well as to selected design parameters. Clear communication of the design to the Contractor will help to identify the appropriate methods of construction and will enable inappropriate methods to be discounted at an early stage.

It is essential that the type of pile to be specified is carefully considered to ensure its suitability in relation to the ground and environmental conditions. Ground conditions must be properly defined by means of adequate site investigation works to permit the appropriate selection of pile type and the proper design of the pile.

Criteria governing the length of piles and testing requirements should be agreed before entering into a contract.

A3.2.3 Ground conditions

A Site Investigation (SI) is a desk study combined with the Ground Investigation (GI).

A GI for geotechnical works should comprise appropriate geotechnical and geo-cnvironmental fieldwork and laboratory testing.

BS EN 1997-1 states that a Ground Investigation Report should comprise all available geotechnical information with a geotechnical evaluation of this information, stating assumptions.

BS EN 1997-1 also states that the Ground Investigation Report shall form a part of the Geotechnical Design Report. The Geotechnical Design Report shall include assumptions, data, calculation methods and results, including a description of the ground conditions, design values of soil and rock properties, statement on risks, and foundation design recommendations.

A planned and well-executed GI, which addresses both design and construction phases, is essential to the successful outcome of any geotechnical project. The Engineer is responsible for advising the Employer on the appropriate scope of the SI for the foundation works, including the adequacy of any provided information. The Employer is responsible for implementing all the Engineer's recommendations.

A comprehensive GI should be carried out in accordance with BS EN 1997-2, the recommendations of the ICE Site Investigation Steering Group publications, and other appropriate guidance documents such as those published by the Association of Geotechnical and Geoenvironmental Specialists (AGS) (see www.ags.org.uk).

A project involving piling requires an appropriate extent of GI to be carried out to assess the choice of pile type, design parameters and constructability, including temporary works. Typical requirements include:

- Several exploration points (BS5930 requires a minimum of three).
- Appropriate geotechnical characterization of the ground which would normally require boreholes.
- Depth of exploration at least to the deepest pile depth plus five times the diameter of that pile or to the depth of the maximum diameter pile plus five times its diameter, whichever is the greater.

A3.2.4 Workmanship

The Contractor's or subcontractor's responsibilities for their workmanship in achieving compliance with the design, drawings and specification provided to them in addition to other relevant information accessible to them, should be stated in the contract or subcontract documents.

Proper supervision of piling and embedded walling works by experienced site personnel is essential. It is preferable that supervision is provided by both the Contractor and the Engineer. The Employer should facilitate supervision by the Engineer. Supervision by the Engineer should be by a competent person with appropriate qualifications and experience.

A3.3 Environment
A3.3.1 Choice of foundation and requirements

The specification of appropriate performance and design requirements is an essential aspect of an economic and sustainable foundation solution. The aim should be to achieve the required foundation performance appropriate for the structural needs. Excessively conservative requirements can substantially increase foundation costs and lead to the inappropriate use of resources and energy.

A3.3.2 Sustainability

Sustainable construction can be defined as development which meets the needs of the present without compromising the ability of future generations to meet their own needs.

Sustainable construction is the set of processes by which a profitable and competitive industry delivers built assets (buildings, structures, supporting infrastructure and their immediate surroundings) which:

- Enhance quality of life and offer customer satisfaction.
- Offer flexibility and the potential to cater for user changes in the future.
- Provide and support desirable natural and social environments.
- Maximize the efficient use of resources.

Building construction and operation can have a significant direct and indirect impact on the environment. Foundation construction is an important part of building construction and it is recommended that the choice of method, the use of appropriate materials, the minimizing of waste (including arisings), and efficient design are all considered to minimize environmental impact.

Carefully planned decisions made early in the design process frequently have a profound influence on the sustainability of the foundation solution. It is important that the Employer, Engineer and Contractor recognize that these decisions must reflect not only the structural requirements of the foundation works but also benefit the local environment and quality of life.

A3.3.3 Design and construction

Environmental aspects to consider during the design and construction of piles or embedded retaining walls include:

- Energy conservation.
- Whole life performance.
- Re-use or removal of existing foundations.
- Noise, transportation and storage of materials during construction and their impact on the surrounding environment.
- Dust and vibration.
- Storage and removal of waste and arisings produced from the construction of piles.
- Re-use of pile arisings.
- Potential contamination of the surrounding ground and groundwater during construction.
- Creation of potential flowpaths between aquifers and contaminated ground.
- Aggressive ground conditions which may cause concrete deterioration or corrosion of steel.

A4 Enquiry documents

Enquiry documents are the documents issued by the Employer stating the basis upon which the Employer would like to receive tenders from the Contractor.

A4.1 Electronic information

The increased use of information technology in the construction industry is having a profound impact not only on communication and design efficiency, but is influencing other areas such as procurement.

It is desirable for the Employer to ensure that the electronic data being supplied is relevant and is clearly indexed allowing the Contractor to be confident that all the necessary information can be accessed efficiently. A protocol for electronic data transfer is provided in Appendix 2.

Information provided on a CD or DVD needs to be clearly indexed. Information provided on an extranet should be designed so that the user receives only targeted updates.

A4.2 Selection of contractors to tender

Tenders should be invited only from Contractors who are suitably experienced with similar foundation works in similar ground conditions.

A4.3 Methods of procurement of contractors

Contractors may be invited to tender for piling or embedded walling works as a main contract to an Employer either under one of the established forms of contract, or under a contract document drawn up especially for management-type contracts.

Where the Employer chooses to name or nominate the subcontractor for the piling or embedded walling works it is normal for a prime cost item to be included in the main contract bills of quantities. It is necessary to ensure that the piling specification and other relevant items from piling subcontract documents are included in the main contract enquiry documents so that tenderers for the main contract are aware of, and can price for, their responsibilities to, and attendance on, the piling subcontractor. Equally the subcontract enquiry documents should include information regarding the form of the main contract, including the appendix to the contract and sufficient detail that the subcontractor is aware of the allocation of responsibilities and liabilities at the time of preparing their tender. Alternatively:

(a) The piling or embedded walling works may be measured and included in the bills of quantities for the main contract, and the document may stipulate that these works shall be executed by any one company from a list of approved piling contractors. The Contractor may have the option of proposing further companies for approval.

(b) The piling or embedded walling works may be measured and included in the bills of quantities for the main contract but without any list of approved piling contractors. In such cases, the Contractor will seek prices from a selection of piling contractors, for whom it is normal practice to seek approval before they are employed on the works.

(c) Following appointment, the Contractor may be instructed to carry out additional works not included in the contract. Should this additional work include piling, tenders for such works should normally be invited from a list of companies agreed by both the Contractor and the Engineer. It is desirable that the enquiry documents should be approved by the Engineer before invitations to tender are issued.

With procedures (a), (b) and (c), the selected piling contractor is normally appointed as a domestic (direct) subcontractor of the Contractor.

A4.3.1 Early contractor involvement

There can be cost and programme benefits for the Employer arising from appointing a Contractor early on in the procurement process. For the negotiation process to be successful for both Employer and Contractor it is important to:

- Involve the Contractor early, but not so early that time is wasted because the project detail is not sufficiently developed for meaningful analysis.
- Use the Contractor's practical and technical expertise.
- Agree budget costs progressively as the design is developed.

- Develop a design programme such that, at its completion, a contract can be signed in sufficient time for the Contractor to organize plant and labour and to order materials.
- Assign responsibilities and eliminate gaps in the scope of work.
- Co-ordinate between subcontractors at interfaces.
- Early Contractor involvement is essential where re-use of existing piles or foundations is envisaged.

A4.4 Instructions to tenderers

Although they will not always form part of a contract, instructions to tenderers should be provided as necessary and should include:

- Any restrictions relating to visits to the site and the name, address and telephone number of any person from whom permission to visit the site must be obtained, should this be a requirement.
- The procedure for submitting a tender and the latest date for its submission.
- The expected date of award of the contract and the likely commencement date of the works.
- The name and address of the main contractor, if appointed, in cases where piling or embedded walling works are to be executed as a named or nominated subcontract.

A4.5 Information to be provided to tenderers

The Engineer should ensure that as a minimum the factual Ground Investigation Report is incorporated in the documents. The Geotechnical Design Report, including the Ground Investigation Report, should also where possible be made available to the Contractor, but may not form part of the contract.

Proposals for piling and embedded walling works can be included within an enquiry for pricing by companies tendering for a main contract. In such circumstances, the tenderers have to obtain quotations from specialist subcontractors during the tender period and difficulties can arise when all relevant geotechnical information, piling specifications, preambles, etc., are not readily identifiable without close scrutiny of voluminous contract documents. Such information should, wherever possible, be collated in a single section within the contract documents, or otherwise identified, so that the companies tendering for the main contract can issue all relevant information in a clear manner to specialist contractors.

The specialist contractor should be provided with the requisite information with sufficient time to enable assessment the documents and, when required, design of the piles or wall elements prior to submission of tender.

A4.6 Measurement/ Outline Bill of Quantities

An outline Bill of Quantities for piling works is given in Appendix 3. This is intended only to illustrate the key items recommended for measurement and is not intended to be comprehensive or applicable for all piling or embedded walling methods. For more detailed information on the measurement of piling or embedded walling works see www.fps.org.uk.

A4.7 Electronic auctions

An increasingly common practice among Employers is the use of electronic auctions. These may extend to the establishment of unpriced but pre-qualified bids based upon a set of documents or, alternatively, may be restricted to qualification based upon very simple criteria. Following such a qualification process each tenderer may then be required to submit on-line a price for their tender on an auction basis whereby the lowest offer would then normally be accepted.

It should be noted that different contractors will have varying practical and technical expertise and the most suitable Contractor may not necessarily be reflected by the cheapest price. Also, piling and embedded walling works involve a great deal of innovation and this fact should be taken into account when evaluating tenders.

A5 Tender documents

Tender documents are the documents that form the Contractor's offer, including those documents referred to within it. These include the enquiry documents unless specifically stated.

A5.1 Information to be supplied by the tenderer

The enquiry documents should give the tenderers the opportunity to provide all of the information required for the proper evaluation of the tenders. In the event that the enquiry documents do not adequately cover the items set out below, the tenderer should include within the offer:

- Validity of tender — the period for which the tender remains open for acceptance, and the period required for execution of the works.
- Insurances — the upper limit to which the tenderer is already insured if it is less than a figure stipulated in the enquiry and will therefore require to be supplemented.
- Working hours — the basis of the tender in respect of the days and hours during which the tenderer expects to work, under normal conditions.
- Method statement — a description of the type of equipment proposed to be used for the execution of the piling or embedded walling works, the method of construction and, where appropriate, any assumptions made regarding the programme and/or sequence of piling operations, the number of piling machines and number of visits allowed.
- Contract period — an estimate of the contract period for working without disruption on the work specified at tender stage.
- Notice — period of notice required for commencement.
- Design period — where applicable, period required for completion of the design after receipt of the 'for construction' information.

A5.2 Working levels

'Working platform level' is the level at which the piling rig stands to carry out the work. Where piling or embedded walling works have to be carried out from a fixed structure or staging, whether permanent or temporary, or where floating or temporary staging may be used at the Contractor's discretion, the bill of quantities should be drafted accordingly.

'Commencing surface' in relation to an item in a bill of quantities is the surface level at the pile or wall element position at the commencement of the boring or driving operation.

Work from any commencing surface which differs from the working platform level and where the Contractor may incur significant additional costs, should be billed separately.

A6 Tender evaluation

Tenders for most civil engineering and building works can be compared equitably by giving primary consideration to the tender sum evaluated from a priced bill of quantities. Piling and embedded walling works are different in that different contractors may have varying practical, technical and unique expertise relevant to the project and the most suitable contractor may not necessarily be reflected by the lowest price.

Specification for piling and embedded retaining walls. Thomas Telford, London, 2007

Pile lengths required for a given load capacity in particular ground conditions will not necessarily be the same for piles of different types and sizes. The lowest tender may not necessarily represent best value for money in terms of the final account after remeasurement. When examining tenders for piling or embedded walling works, the Engineer should consider potential variations in the measured quantities relative to each pile type and any impact on costs.

If the type of pile and/or method of construction or of installation is not specified in the enquiry documents, and is not fully described in the tender submission, then this information should be obtained following receipt of tenders and should be agreed prior to contract award.

If not already having done so, the Contractor should visit and examine the site and its surroundings to satisfy themselves, prior to entering into contract, that the tender offer is appropriate to the available contract documentation and what can be determined from visual inspection.

A6.1 Appointment of a Geotechnical Advisor

For all piling and embedded walling works it is advisable that the Employer appoints a nominated lead professional of Geotechnical Advisor status, as defined by the ICE Site Investigation Steering Group, who can be consulted at all stages of the project through planning, design and construction.

The Geotechnical Advisor should have a key role in the evaluation of tenders. In particular the Geotechnical Advisor should be consulted on tender qualifications and their potential implications on the design and on costs.

The Geotechnical Advisor should have qualifications, experience and expertise appropriate for the foundation works.

A7 Contract documents

Contract documents are the documents that form the contract between the Employer and the Contractor.

A7.1 Documents forming part of a contract

Documents which should form part of a contract and subcontract for piling or embedded walling works are:

(a) Conditions of contract, and of subcontract where applicable, including:
- The location of the site and access thereto.
- Any limitations of working hours.
- Any special conditions limiting noise and vibration or other environmental constraints.
- A detailed programme including any phasing of the works necessitating more than one visit by the Contractor or subcontractor.
- Details of other contractors working on the site at the same time.
- The available working and storage areas, including restrictions on headroom.
- Any known special conditions/restrictions to be imposed.
- Any other appropriate special conditions required by the Employer.

Additionally, in the case of subcontract documents, the following should be incorporated:
- Full details of the relevant conditions of the main contract, notably such matters as insurances, liquidated damages, programme requirements, bonds, etc., which may affect the Contractor's or subcontractor's tender.

(*b*) Enquiry documents including:
- The period for which the tender is to be valid for acceptance.
- Where the Engineer responsible for the piling enquiry documents has not been engaged to supervise the works, it is desirable that this should be so stated.
- Other information such as results of any preliminary pile tests.

(*c*) General and Project Specifications including:
- Allocation of design responsibilities.
- Pile schedules in hardcopy and unlocked electronic format.
- Performance Specification when applicable.
- Wall Manual when applicable (see Clause B1.4.1).

(*d*) Drawings scheduled as contract drawings in the tender document including:
- Details of underground services, existing structures and other known obstructions details.
- Commencing surface levels and/or working platform levels.
- Pile cut-off levels, or, if this information is not available, a statement of the depth of the cut-off levels below the ground surfaces; all levels should be related to an Ordnance datum or other datum as required by the Engineer.

(*e*) Site information and SI data including:
- Geotechnical Design Report including Ground Investigation Report (borehole logs, *in situ* test results, laboratory test results and contamination data).
- Existing ground levels.
- Ground levels at borehole and other test positions relative to the stipulated datum (see (*d*) above) and showing the positions of such boreholes, etc. relative to the outline location of the piling or embedded walling works.
- Electronic GI data in AGS format when available.

(*f*) Bill of quantities based on a tender piling schedule.

(*g*) Form of acceptance.

A7.2 Model clauses suitable for inclusion in contract documents

Where piling or embedded retaining walls are executed under a subcontract, additional provisions must be incorporated in the documents to define the responsibilities of the Contractor and subcontractor. A typical list of attendances and facilities which the subcontractor may require to be provided is presented in Appendix 4 (also see www.fps.org.uk). Some of these clauses may need to be omitted, in whole or in part, to suit the particular needs of the contract works and the subcontract.

Appendix 1

Example Project Specification

B1 SPECIFICATION REQUIREMENTS FOR PILING AND EMBEDDED RETAINING WALLS

B1.2 Project Specification

(a) Role of the Engineer
As stated in the contract documents.

(b) Location and description of the site
The site is located in 'anytown' (National Grid Reference: XX).

  The site is bounded to the east by a road, to the west by a hill, and to the north and south by existing industrial buildings. The site is presently occupied by a two-storey building, which is to be demolished before the Works commence.

(c) Nature of the Works
The Works comprise the construction of 101 piled foundations, but not the pile caps, for a new four-storey building.

(d) Working area
As stated in the contract documents.

(e) Sequence of the Works and other works proceeding at the same time
Sequence as stated in the contract documents. Pile probing is to be carried out by others before the Works commence.

  No other activities on site when the Works are being carried out.

(f) Contract drawings
The Piling Drawing is drawing No. XX999 Pile Details.

(g) Office and other facilities for the Engineer
As stated in the contract documents.

(h) Submission of information (in addition to Table B1.1)
The information which shall be submitted at the stated stages of the Works Contract is listed in Table B1.1 of the Piling Specification.

  In addition, the Tenderer shall submit the following details with his tender:

  XXX

(i) Responsibility for design, including any division of this responsibility
Option 2 — Contractor design of piles is specified, see Clause B1.4 of the Piling Specification.

Design responsibility	Engineer	Contractor
1. Design of foundation scheme	×	
2. Choice of piling or walling method		×
3. Design of piles or wall elements to carry Specified loadings		×

(j) Design standards and criteria for the piles or walls including design life
Pile design to be in accordance with British Standard XX and pile design life to be 50 years.

(k) Constraints on design
None.

(l) Working platform and commencing surface level
Working platform and commencing surface level will be at +20.0 mOD.

(m) A schedule of Specified working loads or Representative actions
Pile safe working load = 750 kN.

(n) Pile or wall element dimensions
All pile sizes acceptable.

(o) Preliminary piles and trial bores/drives/panels
To be determined by Contractor.

(p) Performance criteria for the structure to be supported on the piles or by the wall
The design of the superstructure assumes that the settlement of any part of the structure at working load will not exceed 20 mm and that the differential settlement between adjacent columns at working load will not exceed 10 mm.

(q) Performance criteria for piles under test or wall elements during service (see Tables B1.2, B1.3)
The Contractor will determine if pile tests will be carried out on this site to comply with his design. Table B1.2 states the performance criteria expected of the piles.

Pile ref.	Permitted type(s) — performance specification Section No.	Specified working load SWL (kN)	Pile designation	Minimum factor of safety	Design verification load DVL (kN)	Permitted settlement at DVL (mm)	Permitted settlement at DVL $+\frac{1}{2}$ SWL (mm)	Minimum pile length from cut-off level to toe (m)	Minimum or maximum pile diameter or dimensions of cross-section (mm)
P1 to P101	All types permitted	750	To be confirmed	2	750	10	20	None	None

(r) Sampling and testing of materials (other than concrete)
Concrete strength and consistence shall be measured as chapter B.19 of the Piling Specification.

(s) Permissible damage criteria for existing critical structures or services
No damage is to be caused to neighbouring structures or services. If the Contractor is of the opinion that by any means arising from the Works neighbouring property may be damaged this shall be stated in his tender and the extent and nature of the necessary protective or other temporary works described therein.

(t) Additional temporary works plant and duration of loading for which the working platform should be designed
The working platform shall be designed for loadings from piling plant as supplied by the Contractor, for the duration of the Works.

The working platform is also to be designed for all handling cranes, mobile cranes and other ancillary plant for loadings supplied by the Contractor.

(u) Site datum and site grid
As stated in the contract documents.

(v) Restrictions on permissible working hours
As stated in the contract documents.

(w) Restrictions on noise and vibration levels
As stated in the contract documents.

(x) Site Investigation including geotechnical and geo-environmental information, and the need for further Site Investigation

The following Ground Investigation Report is included:

Bloggs SI (2006) 'Mysite Factual Ground Investigation Report'

No further SI is planned.

(y) Disposal of excavated material and trimmed excess pile and wall material

Spoil generated from the site will require disposal in accordance with the Environmental Protection Act 1990, the Landfill (England and Wales) Regulations 2002, the Waste Management Licensing Regulations 1994 and any other relevant statutory instrument and guidance.

(z) Other particular technical requirements

As stated in the contract documents. The Contractor is required to work to a Quality Management system established in accordance with BS EN ISO 9001:2000. Details shall be provided prior to commencement of work on site.

Appendix 2

Recommended electronic tendering protocol
Electronic tendering protocol for geotechnical works

- All electronic information provided on an extranet shall be properly indexed and organized with all information relevant to the geotechnical works being easily identifiable.

- All electronic information provided on a CD or DVD, or similar, shall be properly indexed and organized with all information relevant to the geotechnical works being easily identifiable.

- It is preferable that only information specifically relating to the geotechnical works is provided.

- All electronic information shall be provided in *.pdf format that is easy to read and locked so that no unauthorized amendments can be made. This information may include:

  - Contract conditions
  - Specifications and schedules
  - Site investigation reports, including the borehole logs
  - Drawings.

- Drawings shall also be provided unlocked in AutoCAD *.dwg format.

- Relevant forms shall also be provided unlocked in MS Word or Excel format.

- All schedules shall also be provided unlocked in MS Excel format (see www.fps.org.uk).

- Ground Investigation data shall also be provided in AGS format pre-approved by the Client.

Specification for piling and embedded retaining walls. Thomas Telford, London, 2007

Appendix 3

Example Bill of Quantities

Reference	Item description	Unit	Quantity	Rate	Amount £ P
A	Transport piling plant and equipment to site, set up, dismantle and remove upon completion. Per visit	sum			
B	Move between area * and area *	sum			
	*The following work to be carried out from a working platform level of *mOD with a Commencing Surface level of *mOD*				
C	Movement of plant and equipment to each pile position, including setting up rig *mm specified diameter *mm specified diameter	nr nr			
D	Bore pile shafts to depths not exceeding *m *mm specified diameter *mm specified diameter	m m			
E	Concrete quality *in pile shafts *mm specified diameter *mm specified diameter	m m			
F	Enlarged bases including forming and concrete quality *therein Shaft dia. Base dia. *mm *mm *mm *mm	nr nr			
G	Grade 500 straight bars in reinforcing cage *mm diameter *mm diameter	t t			
H	Grade 500 helical binding *mm diameter at *mm pitch	t			
I	Backfill open shafts with *	m³			
J	Remove excavated material from the area around the pile position during boring operations, including loading and depositing *	m³			
K	Cut down concrete pile shaft to the specified cut-off level, prepare exposed head and reinforcement to receive capping: load and dispose of debris *mm specified diameter *mm specified diameter	nr nr			
L	Make concrete cube and test	nr			
M	Transport equipment to site for pile load tests not exceeding *kN and remove upon completion of tests	sum			
N	Pile load test on a *mm diameter working pile to *kN, including reaction piles	nr			
O	Cut down *mm pile cap and tops of reaction piles to a depth of *m, dispose of debris, backfill, and make good disturbed site surface	nr			
P	Carry out approved integrity tests (minimum 20 per visit) and provide report by approved specialist	nr			
Q	Remove obstruction in bore, utilizing standard equipment on site. Each plant and gang	h			
R	Charge for plant and labour delayed by causes beyond the piling contractor's control or when instructed. Each plant and gang	h			
S	Design fee	Sum			
T	Design, construct, maintain and repair Working Platform	Sum			

Appendix 4

Typical Schedule of Attendances and Facilities to be provided by the Client to a Piling or Embedded Walling Specialist, including Working Platform Certificate

COMPANY NAME: _____

CONTRACT NAME: _____

TENDER REF.: _____

<u>SCHEDULE OF ATTENDANCES AND FACILITIES TO BE PROVIDED BY THE CLIENT TO A PILING OR EMBEDDED WALLING SPECIALIST</u>

For the purposes of this document the following definitions shall apply:

Specialist	— Piling/Embedded Walling Specialist
Client	— Person directly employing the Specialist
Specialist Works	— Any operation or installation undertaken by the Specialist
Working Surfaces	— Any platform, ramp, lay-down, storage area or the like required to be used by the Specialist

The following attendances and facilities shall be provided and maintained at all times (including additional working hours if necessary) for the duration of and in relation to the Specialist Works, free of charge to the Specialist and in a manner so as not to disrupt or restrict the regular progress of the Specialist Works.

1. **Notices and Approvals.** Giving all notices and obtaining all necessary approvals, licences and sanctions, including but not limited to any method statement and/or design submission approvals, planning consents, party wall agreements, wayleaves, easements, possessions, rights of way or access and the like from third parties.

2. **Rates and Fees.** Payment of any rates or fees which may become payable due to occupation of the Specialist Works.

3. **Design Information.** No later than three working weeks (or other period as agreed) in advance of the programmed commencement of the Specialist Works on site, provide the Specialist with complete and final construction issue drawing(s), specification(s), pile/panel load schedule(s) in electronic spreadsheet format, and other relevant information required to undertake the Specialist Works. Any variation to these documents to be notified to the agreed contact in hardcopy format.

4. **Existing Services.** Accurate setting out and clear and robust marking or exposing on site the exact location of existing underground/overhead services and providing a drawing on which their positions in line and level are accurately plotted relative to the Specialist Works. Protection, diversion or removal of such services to prevent damage from the Specialist Works. The location and sealing off of all disused pipes or ducts in order to prevent the entry of concrete, grout, slurry or drilling fluids during construction. Subsequent to the foregoing, issue to the Specialist's authorized representative on site a permit to dig or similar system on a daily or as otherwise agreed basis.

5. **Obstructions.** Prior removal of overhead, surface or underground obstructions which may impede the Specialist Works and backfilling of excavations and voids with a suitable material which will not obstruct or be deleterious to the works but which will ensure the stability of the Specialist Works and will maintain compliance with Items 11, 12 and 13.

6. **Shoring/Underpinning.** Shoring and underpinning as necessary, including the removal, replacement or adjustment of timbering or shoring which may impede the Specialist Works.

Specification for piling and embedded retaining walls. Thomas Telford, London, 2007

7. **Clearance.** The provision of clearance around working positions and protection to adjacent works, structures (including underground structures), third party property and other site boundary constraints, to allow the execution of the Specialist Works.

8. **Fencing, Hoardings, etc.** Hoardings, fences, noise and splash barriers, statutory warnings, signage, flagmen or the like as necessary to protect the works, plant, materials, personnel and to provide protection to third party property and members of the general public at or adjacent to the site boundaries. This shall include protection from concrete/slurry splashing, exhaust, oil, grease etc.

9. **Traffic.** Control or diversion of road, rail or water borne traffic, including all necessary arrangements and payment of charges in connection with any road closures, lane rental and/ or suspension of parking bays, rail possessions and the like.

10. **Watching.** Provision of security to safeguard the plant, equipment, materials on the site and the Specialist Works.

11. **Access.** Full, free, controlled and uninterrupted access onto the site(s) from hard road to Working Surfaces designed, installed, maintained and repaired as required under Item 13, including protective mats and all other equipment and measures necessary to ensure the safety of pedestrians and to avoid the risk of damage to third party property, including road surfaces, kerbs, footpaths/pavements, surfaces and services. Such access to facilitate the safe erection, operation and movement on, off and around the site(s), to and between lay-down, working and storage areas, pile/panel positions and test piles/panels for plant including provision and where necessary relocation of ramps, including to concrete agitators, to a gradient not steeper than 1 in 10.

12. **Hardstanding and Storage Areas.** Provision and subsequent removal of conveniently situated hardstanding and storage areas designed, installed, maintained and repaired as required under Item 13 to facilitate the safe operation and erection of plant and equipment, storage of plant, equipment and materials, offices, sheds and the manufacture of reinforcing cages.

13. **Working Platform.** Design, installation, maintenance, repair and subsequent removal of free draining Working Surfaces in accordance with the requirements of the attached Working Platform Certificate, a signed copy of which must be provided to the Specialist prior to the Specialist's commencement. The design(s) shall include for all piling rigs, ancillary plant and equipment and wheeled transport including articulated and ready mixed concrete lorries and the checking of any retaining structures which support the Working Surfaces.

14. **Surface Water and Groundwater.** Any pumping or drainage required to keep the site free of flooding, surface water or any water and slurry arising from the Specialist Works.

15. **Health, Safety and Welfare.** Health, safety and welfare facilities as required to comply with the Construction (Health, Safety and Welfare) Regulations 1996 in particular Regulation 22 and Schedule 6 and any other statutory regulations or rules, orders or regulations of any authority having powers related to the Specialist Works. Such facilities as a minimum to include the supply of drinking water, washing facilities including hot and cold running water, male and female changing/drying and toilet facilities.

16. **Environmental.** Attendances, facilities and licences to comply with environmental legislation and rules/order or regulations of the Environment Agency and other statutory bodies.

17. **Temporary Lighting.** Background and task lighting to Working Surfaces to allow safe working and safe access and egress and to facilitate execution of the Specialist Works.

18. **Water Supply.** Within the working, storage and preparatory operation areas, potable water supply at mains pressure take-off points and sufficient for the operations, including (where applicable) charging of bores, bentonite/drilling fluid mixing, concrete/grout mixing and cleaning of plant.

19. **Electricity.** Within the working, storage and preparatory operations areas, sufficient power take-off points and power.

20. **Communication Facilities.** Provision of site telephone facilities and/or electronic data transfer facilities and/or designated areas for the use of mobile telephones as applicable.

21. **Guide Walls.** The setting out and checking, construction and later removal of reinforced concrete guide walls (where required). The top of the guide wall shall be a minimum of 1.5 m above water levels encountered during construction of the Specialist Works.

22. **Setting Out.** Provision of clear, accurate and robust setting out and levels, checking and maintenance of individual pile/panels positions as necessary throughout the contract, provision of permanent datum points, base lines, structural grid lines as required and as built survey information.

23. **Checking.** Checking the positions and levels of all piles/panels, during the progress and on completion of the Specialist Works (where practicable) including the prompt counter-signature of the Specialist's record sheets by an authorized representative of the Client before the Specialist's plant has left site.

24. **Attendant Excavator.** The provision of a 360 degree tracked excavator (and appropriately qualified operator) with a certificated minimum 1 tonne lift capacity in full time attendance to each rig for all operational purposes including removal of excavated or displaced material (including mud slurry and excess concrete from around the Specialist Works) in sufficient time to prevent the formation of spoil heaps impeding the Specialist Works, the handling of concrete delivery pipes, the unloading, transportation and installation of reinforcement and other general duties.

25. **Spoil Removal and Waste Management.** Classification, removal from site and disposal of excavated or displaced material including mud slurry and excess concrete from the Specialist Works in accordance with current legislation, rules/order or regulations of the Environment Agency and other statutory bodies including notification to the Environment Agency where arisings are classified as hazardous waste. Classification and disposal from site of waste water from the Specialist Works (such as concrete wash out from agitators and ready-mixed concrete delivery trucks) to comply with current legislation, including for filtering or the provision of settlement tanks as appropriate.

26. **Wheel and Road Cleaning.** Manned wheel-cleaning facilities and/or road cleaning, as necessary.

27. **Protection.** Protection of the works where taken over by other trades or contractors or where the Specialist has left site, whichever occurs first.

28. **Empty Bore/Panel Reinstatement.** Backfilling of any empty bore/panel excavation with a suitable material which will not obstruct or be deleterious to the works but which will ensure the stability of the Specialist Works and will maintain compliance with Items 11, 12 and 13.

29. **Building up.** Building up of piles/panels or any necessary modification to the sub-structure as a result of the piling platform level being less than 0.45 m (or as otherwise agreed) above the specified cut-off level.

30. **Trimming.** Cutting back heads of piles/panels, including overbreak and casing left in, down to the specified cut-off levels (including preparation of the surface of the pile head for integrity testing); cutting back overbreak from the face of retaining walls; disposing of debris; and preparing, straightening, cutting and bending the reinforcement, including test piles/panels, caps and reaction piles/panels or anchors; provision of all necessary engineering, setting out and level control to facilitate the installation of debonding materials/agents to agreed pile/panel reinforcement placement tolerances. Stripping formwork to recesses and box outs etc., including templates to starter bars or bolts.

Working Platform Certificate

Project Name	
Section/Activity	

Part 1 — WORKING PLATFORM DESIGN

Equipment to be used on site	
Maximum plant loading	

(Note: 'Working Platforms for Tracked Plant: Good practice guide to the design, installation, maintenance and repair of ground-supported platforms' is available from BRE Bookshop — Tel. 01923 664000.)

Part 2 — WORKING PLATFORM INSTALLATION

The Working Platform on the above stated work site has been designed and installed to safely support the equipment detailed on this certificate and it will be maintained and repaired, and reinstated to the as installed condition after any excavation or damage, throughout the period when the equipment is on the site.

Signature		Name	
Position	Principal Contractor	Date	
Organization		Address	

A completed copy of this certificate signed by the Principal Contractor must be given to each user of the working platform prior to commencement of any works on site.

> The HSE has worked closely with the FPS to develop this initiative and supports the principle of reducing accidents by the certification of properly designed, prepared and maintained working platforms.

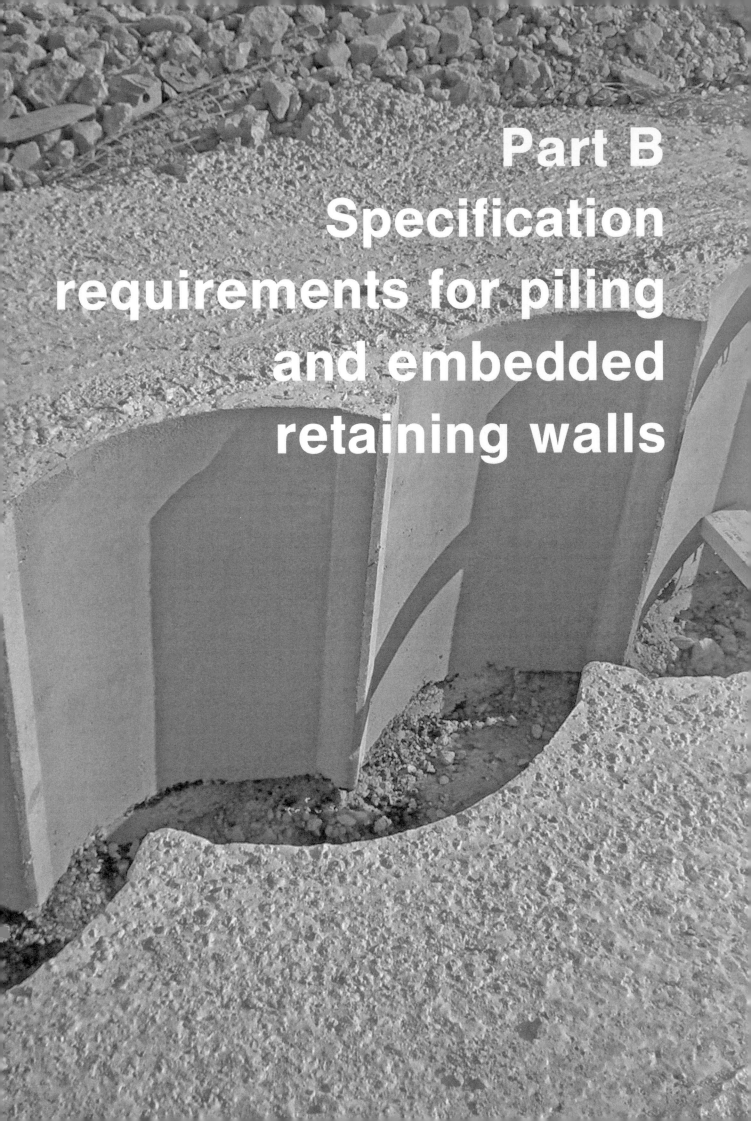

Part B
Specification requirements for piling and embedded retaining walls

Specification requirements

B1 Specification requirements for piling and embedded retaining walls

B1.1 Standards

All materials and workmanship shall be in accordance with the appropriate British Standards, European Standards, Codes of Practice and other specified standards current at the date of tender.

Where there is any conflict of requirements in this Specification, the requirements of the Project Specification shall take precedence. Where there is any conflict of requirements of this Specification with any published standard, this Specification shall take precedence over that standard, but only for the part of the Works to which the conflicting Specification applies.

B1.2 Project Specification

The following matters are, where appropriate, described in the Project Specification:

(a) role of the Engineer

(b) location and description of the site

(c) nature of the Works

(d) working area

(e) sequence of the Works and other works proceeding at the same time

(f) contract drawings

(g) office and other facilities for the Engineer

(h) submission of information (in addition to Table B1.1)

(i) responsibility for design, including any division of this responsibility

(j) design standards and criteria for the piles or walls including design life

(k) constraints on design

(l) working platform and commencing surface level

(m) a schedule of Specified Working Loads or Representative actions

(n) pile or wall element dimensions

(o) preliminary piles and trial bores/drives/panels

(p) performance criteria for the structure to be supported on the piles or by the wall

(q) performance criteria for piles under test or wall elements during service (see Tables B1.2, B1.3)

(r) sampling and testing of materials (other than concrete)

(s) permissible damage criteria for existing critical structures or services

(t) additional temporary works plant and duration of loading for which the working platform should be designed

(u) site datum and site grid

(v) restrictions on permissible working hours

(w) restrictions on noise and vibration levels

(x) Site Investigation including geotechnical and geo-environmental information, and the need for further Site Investigation

(y) disposal of excavated material and trimmed excess pile and wall material

(z) other particular technical requirements

Items (aa) to (gg) are additional particular requirements for walls only:

(*aa*) special requirements for retaining walls

(*bb*) loads on walls and excavation depths

(*cc*) water retention function and degree of retention for retaining wall in both temporary and permanent conditions

(*dd*) design external groundwater level for the water retention system and for watertightness inspections

(*ee*) design water retention Grade for permanent overall system and the role of the retaining wall as a system component

(*ff*) programme for watertightness inspections

(*gg*) permissible limits on ground movements during wall installation

Table B1.1 Submission of information. (The following submissions shall be made by the Contractor to the Engineer at the time stated and in accordance with the Project Specification. Detailed requirements are listed under the Clause number indicated in the relevant position.)

Section	Item	At Tender	Prior to commencing the Works	During the Works
B1	Progress Report			B1.3 (first day of each week)
	Design option 1 — confirmation of method, equipment and programme	B1.4		
	Design option 1 — Wall Manual	B1.4.1 (for walls only)		
	Design option 2 — preliminary pile schedule	B1.4		
	Design option 2 — design and numbered pile schedule		B1.4 (1 week before)	
	Design option 3 — Wall Manual	B1.4 and B1.4.1		
	Design option 3 — preliminary wall schedule	B1.4		
	Design option 3 — design and numbered wall schedule		B1.4 (1 week before)	
	Changes in ground conditions			B1.7
	Method for removal of soil and debris from face of a wall		B1.9.1 (for walls only)	
	Method for repair of a wall		B1.9.2 (for walls only)	
	Method of construction, plant and monitoring	B1.10	B1.10 (1 week before)	
	Methods for dealing with man-made and natural obstructions and voids	B1.10.1		
	Construction programme	B1.11 (provisional)	B1.11 (detailed)	B1.11 (daily)
	Records			B1.12 (within 24 hours)
	Plans for Surveys and monitoring of movements		B1.13.2	
	Risk of damage			B1.13.2 (immediately)
	Piling/walling sequence		B1.13.3	B1.13.3
	Uplift of driven piles			B1.14.4 (once per week)
	Laterally displaced driven piles			B1.14.4
	Confirmation of full time supervision	B1.15		
	Supervisor CV		B1.15	
	Quality Plan		B1.15 (1 week before)	
	Non-conformances			B1.16 (within 7 days of discovery)
	Method of Head preparation	B1.17		

Specification for piling and embedded retaining walls. Thomas Telford, London, 2007

Table B1.1 continued

Section	Item	At Tender	Prior to commencing the Works	During the Works
B2	Variations in length compared to design length			B2.4.7 (within 24 hours)
B3	Temporary casing details (if to be used to core through obstructions or adjacent piles for a secant pile wall)	B3.3.2		
	Procedure for reinforcement		B3.4.1 (1 week before)	
	Details of permanent casing, installation and backfilling of annulus	B3.5.1		
	Loss of support fluid			B3.5.3 (within 24 hours)
	Time for pile construction greater than 12 hours	B3.5.5		
	Annulus formed during casing installation	B3.5.5		
	Previous experience with underream tool and details	B3.5.8.2		
	Method of grouting	B3.5.11.2		
	Measurement system for pile uplift during base grouting	B3.5.11.4		
	Grouting records			B3.6.2 (within 24 hours)
B4	Rig operator CV		B4.4	
	Re-augering			B4.4.1 (within 24 hours)
	Variations in length compared to design length			B4.4.4 (within 24 hours)
	For DA piles the procedure to initiate concrete delivery	B4.4.5.2		
	For DA piles details of reversal of auger rotation during extraction	B4.4.5.3		
	Automatic and manual monitoring system	B4.4.9		
	Target boring and concreting parameters		B4.4.9.3	
	Calibration certificates		B4.4.9.4	
	Additional records			B4.5.1 (within 48 hours)
B5	Variations in length compared to design length			B5.4.2 (within 24 hours)
	Enlarged pile base	B5.4.5		
B6	Inspection		B6.3.2 (1 week notice before manufacture)	
	Test certificates		B6.3.2	
	Preliminary pile results		B6.4.1 (1 week before ordering)	
	Steel screw pile procedure	B6.4.4.1		
	Variations in length compared to design length			B6.4.8 (within 24 hours)
	Welding procedures and electrodes	B6.6		
	Welder proficiency		B6.6.1 (1 week before)	
	Fabrication details, drawings and welding procedures		B6.6.2 (1 week before fabrication)	

Section	Item	At Tender	Prior to commencing the Works	During the Works
	Non-destructive testing of welds		B6.6.3 (1 week after testing)	B6.6.5.3 (1 week after testing)
B7	Certification of timber		B7.3.5	
	Inspection		B7.4.1	
	Certificates of treatment		B7.4.2	
	Changes to supplied length			B7.4.6 (within 24 hours)
	Departure of true alignment of spliced piles			B7.4.7 (within 24 hours)
B8	Alternatives to steel reinforcement	B8.3.4		
	Stability of excavation if specified panel dimensions too large	B8.4.6		
	Construction drawings		B8.5.1	
	Loss of support fluid			B8.5.3 (before proceeding)
	Type of stop-end and whether temporary or permanent	B8.5.6		
	Additional concreting records			B8.6.1 (within 24 hours)
	Additional records of special measures			B8.6.2 (within 48 hours)
B9	Details of mix for primary piles		B9.3.1	
	Alternatives to steel reinforcement	B9.3.3		
	Demonstration if for temporary works a guide wall is not deemed necessary		B9.5.1	
B10	Alternatives to steel reinforcement	B10.3.2		
B11	Details of structural steel section and calculations		B11.3.1 at Tender stage	
	Method of handling and placing steel section	B11.5.3		
B12	Details of alternative fabricated piles to that specified		B12.3.2	
	Test certificates		B12.3.3	
	Inspection		B12.3.3 (1 week notice before manufacture)	
	Brand and properties of interlock sealant	B12.3.5		
	For temporary works, specification for sealing interlocks	B12.3.5		
	Methods for access, handling, pitching, guiding and installing	B12.4.4		
	Details of pre-augering or jetting	B12.4.4, B12.4.9		
	Declutching, damage or separation of sheetpiles			B12.4.4
	Pile refusal or pile damage			B12.4.4
	Variations in length compared to design length			B12.4.8 (within 24 hours)
	Methods of assisting pile installation	B12.4.9		
	Corrosion protection	B12.5		
	Welding procedures and electrodes	B12.6		

Table B1.1 continued

Section	Item	At Tender	Prior to commencing the Works	During the Works
	Welder proficiency		B12.6.1 (1 week before)	
	Procedures for seal welds			B12.6.2 (1 week before seal welding)
	Details of corner fabrication where creep or shrinkage has occurred			B12.6.3
	Non-destructive testing of welds			B12.6.4 (1 week after testing)
B13	Specialist integrity testing contractor		B13.6	
	Integrity testing equipment, method statement and programme		B13.6	
	Preliminary integrity test results			B13.8 (within 24 hours)
	Integrity testing report			B13.8 (within 10 days of testing)
	Anomalous integrity test results			B13.9 (within 24 hours)
B14	Notice of construction of preliminary pile		B14.3.1 (at least 48 hours before construction)	
	Notice of construction of working pile		B14.4.1 (at least 48 hours before construction)	
	Design and construction of load application system			B14.1 (before testing)
	Boring or driving records			B14.3.3 (within 24 hours)
	Calibration certificates			B14.6 (before test commences)
	Commencement of test		B14.8 (48 hours before commencement of test)	
	Preliminary results			B14.10.1 (within 24 hours)
	Pile test report			B14.10.1 (within 10 days of testing)
B15	Details of loading, measurement and reaction system			B15.1 (1 week before)
	Inspection of preliminary pile or working pile test			B15.1 (48 hours before testing)
	Construction of a preliminary pile		B15.3.1 (48 hours before)	
	Preliminary pile materials		B15.3.2	
	Preliminary pile boring or driving record			B15.3.3 (within 24 hours)
	Use of working piles as reaction piles		B15.9.4	
	Integrity test results of working piles used as reaction piles			B15.9.4 (testing within 48 hours, results within additional 24 hours)
	Calibration certificates			B15.10.2 (48 hours before commencement of test)

Section	Item	At Tender	Prior to commencing the Works	During the Works
	Preliminary results			B15.14.1 (within 24 hours)
	Pile test report			B15.14.1 (within 10 days of testing)
	Pile test interpretation			B15.15
B16	Proposed method		B16.1	
	Coating material		B16.3.1 (1 week before application)	
	Details of sleeving		B16.4.1 (1 week before)	
B17	Specialist instrumentation supplier and personnel		B17.1, B17.4.1	
	Inclinometer base readings			B17.3.2.1
	Datum for precise levelling		B17.3.6.1	
	Location and precision of instruments and targets		B17.3.6.2	
	Calibration Certificates		B17.5.2 B17.5.3	
	Report			B17.6
B18	Support fluid details	B18.1		
	Evidence of use		B18.3 (14 days before)	
	Records of compliance testing			B18.6
	Method, frequency and locations of sampling and testing		B18.6	
B19	Mix composition		B19.2.2	
	Consistence		B19.2.4	
	Evidence of ASR compliance		B19.2.5	
	Details of cement type and additions	B19.4.2.1		
	Source of supply of aggregates		B19.4.2.2	
	Water tests		B19.4.2.3	B19.4.2.3
	Consistence testing			B19.8.2
	Strength testing of concrete			B19.8.4
	Details of spacers		B19.9.4	
	Strength testing of structural grout			B19.10.5

B1.3 Progress report

During the Works the Contractor shall submit to the Engineer on the first day of each week a progress report showing the current rate of progress and progress during the previous period on all important items of each section of the Works.

B1.4 Design

The Contractor's obligations and liabilities in respect of the construction and the design shall be fully set out in the contract documents.

Option 1 — Engineer design of piles and walls
The Contractor shall construct piles or wall elements of the type(s) and dimensions, having the qualities of materials and workmanship, and in the locations specified. Before the commencement of installation of the working piles, the Engineer shall provide the Contractor

with a numbered schedule of pile or wall element sizes and lengths. For walls, assumptions relating to the wall propping system and proposed construction sequence shall be listed by the Engineer.

The Contractor shall confirm in the tender that the method of working, equipment and programme are compatible with the installation of the piles or wall elements to the required penetrations in the ground conditions indicated by the results of the Site Investigation, and with the stated design parameters used. The reliance being placed upon the Contractor shall be deemed to be in respect of the Contractor's skill, care and diligence as an experienced contractor executing workmanship and not in respect of any design expertise the Contractor may possess.

Option 2 — Contractor design of piles

The pile locations and Specified Working Loads (or Representative Actions) shall be provided in the Project Specification. The specified limits for load-settlement behaviour of individual piles shall be stated in Table B1.2 of the Performance Specification.

The Contractor shall design and construct the piles having the qualities of materials and workmanship specified and which upon testing meet the requirements of the Performance Specification. The Contractor's design shall comprise the calculation of individual pile sizes and lengths based on the ground conditions revealed by the Site Investigation to carry the Specified Working Loads within the specified limits for load-settlement behaviour in Table B1.2.

The Contractor shall provide with the tender a preliminary pile schedule of sizes and lengths and their corresponding allowable capacities to meet the requirements of the Specification. Other design details submitted at tender stage by the Contractor shall be as stipulated in the Project Specification.

The Contractor shall submit to the Engineer one week before the commencement of the Works the design and a numbered schedule of sizes and lengths of the piles including reinforcement and their corresponding allowable capacities to meet the requirements of the Specification.

Where the Contractor's design uses a variation in the location of the piles provided by the Engineer, the Contractor shall provide the information for the setting out of the individual piles.

Option 3 — Contractor design of wall elements

The location of the wall and, if applicable, the Specified Working Loads shall be provided in the Project Specification. The permitted

Table B1.2 Performance Specification criteria for piles

Pile ref.	Permitted type(s) — performance specification Section No.	Specified working load SWL (kN)	Pile designation	Minimum factor of safety	Design verification load DVL (kN)	Permitted settlement at DVL (mm)	Permitted settlement at DVL $+\frac{1}{2}$ SWL (mm)	Minimum pile length from cut-off level to toe (m)	Minimum or maximum pile diameter or dimensions of cross-section (mm)

Table B1.3 Performance Specification criteria for wall elements

Wall ref.	Permitted type(s) — Performance Specification Section No.	Maximum excavation level	Temporary propping details	Permanent propping details	Construction sequence	Constraints to the wall	Permitted lateral wall deflection (mm)	Watertightness criteria/minimum depth below excavation (m)	Minimum or maximum wall element diameter or dimensions of cross-section (mm)

limits for lateral wall deflection shall be stated in Table B1.3 of the Performance Specification. All constraints to the Contractor's design shall be specified in Table B1.3.

The Contractor shall state all necessary assumptions that have been made in the design in the Wall Manual together with an allocation of responsibilities for verifying these assumptions. The Contractor shall provide the Wall Manual with the tender.

The Contractor shall design and construct the specified walls having the qualities of materials and workmanship specified. The Contractor's design shall be compatible with the specified requirements of the Performance Specification. The Contractor's design shall include justification showing that the expected movements are within the specified limits in the Performance Specification. The Contractor's design shall comprise the calculation of wall element sizes and lengths based on the ground conditions revealed by the Ground Investigation.

The Contractor shall provide with the tender a preliminary wall schedule of sizes and lengths to meet the requirements of the Specification. Other design details submitted at tender stage by the Contractor shall be as stipulated in the Project Specification.

The Contractor shall submit to the Engineer one week before the commencement of the Works the design and a numbered schedule of sizes and lengths of the wall elements including reinforcement and their corresponding allowable capacities to meet the requirements of the Specification.

B1.4.1 Wall Manual

The Contractor shall provide with the tender a Wall Manual.

For Option 1 the Wall Manual shall state the wall propping system and construction sequence that is proposed.

For Option 3 the Wall Manual shall confirm that the Contractor understands the constraints to the design and shall state all necessary assumptions together with an allocation of responsibilities for verifying these assumptions. The Wall Manual shall confirm, or otherwise, that the permitted lateral wall deflection performance stated in Table B1.3 can be achieved and shall state the wall propping system and construction sequence that is proposed.

The Wall Manual, as a minimum, shall comprise a design summary and a construction sequence drawing. The Contractor shall ensure that the Wall Manual forms part of the contract for the excavation and substructure works.

B1.5 Materials

The sources of supply of materials submitted under the requirements of Table B1.1 shall not be changed until the Contractor has demonstrated that the materials from the new source can meet all the requirements of the Specification.

Materials failing to comply with the Specification shall be removed promptly from the site.

B1.6 Safety
B1.6.1 Standards

Safety precautions shall comply with all current legislation.

B1.6.2 Working platform

The Contractor shall design, construct, maintain and repair, for the duration of piling and testing operations, a working platform of sufficient strength and thickness for all plant proposed for use on the site, and for any other plant listed in the Project Specification.

B1.7 Ground conditions

No responsibility is accepted by the Engineer or Employer for any opinions or conclusions given in any factual or interpretative Site Investigation reports.

The Contractor shall report immediately to the Engineer any circumstance which indicates that in the Contractor's opinion the ground conditions differ from those reported in or which could have been inferred from the Site Investigation reports, Geotechnical Design Report, or from preliminary pile results or from trial bores, drives or panels.

B1.8 Installation tolerances

Table B1.4 provides standard installation tolerances for piles and embedded retaining walls. If alternative tolerances are considered necessary then these shall be stated in the Project Specification.

B1.8.1 Setting out

Marker pins for the pile or wall element positions shall be set out and installed by the Contractor. Immediately prior to installation of the piles or wall elements, the pile or wall element positions shall be checked by the Contractor.

The position and verticality/alignment of any auger, tool, casing and/or liner shall be checked by the Contractor immediately prior to installation.

For piles or wall elements with casings and/or liners, the Contractor shall check the position of the casing or liner during and immediately after placing.

Any checks by the Engineer shall not relieve the Contractor of his responsibility.

Table B1.4 Standard installation tolerances

Tolerance	All bearing pile types	All embedded retaining wall types constructed without a guide wall	All embedded retaining wall types constructed with a guide wall
Plan position for piles/walls with cut-off level above or at commencing surface	75 mm in any direction	75 mm	25 mm
Plan position for piles/walls with cut-off level below commencing surface	75 mm in any direction plus additional tolerance in accordance with rake and vertical deviation below	75 mm + 13.3 mm for every 1 m below cut-off level	25 mm + 13.3 mm for every 1 m below cut-off level
Maximum permitted deviation of the finished pile/wall element from the vertical at any level	1 in 75 at any level	1 in 75 for exposed face	1 in 75 for exposed face
Maximum permitted deviation of raked piles	Rake <1 in 6; 1 in 25 Rake >1 in 6; 1 in 15	n/a	n/a

B1.8.2 Forcible corrections to piles

Forcible corrections to concrete piles to overcome errors of position or alignment shall not be made. Forcible corrections to steel piles to overcome errors of position or alignment shall not be made, unless in exceptional circumstances, and only with the permission of the Engineer.

If forcible corrections are made to steel piles the Contractor shall demonstrate that the integrity, durability and performance of the piles have not been adversely affected.

B1.9 Waterproofing of retaining walls

The embedded retaining wall shall be designed for the degree of temporary and/or permanent water retention for an external ground-water level, as stated in the Project Specification. This clause refers to the wall and the joints between wall elements only.

Waterproofing measures shall be able to accommodate lateral wall deflections as described in Table B1.3, measured relative to the as-built position of the wall prior to excavation.

Temporary case: where required in the Project Specification, for all stages in the excavation sequence, the retaining wall shall provide the degree of water retention stated in the Project Specification.

Permanent case: where required in the Project Specification, the retaining wall shall be used as part of the overall protection system as defined by BS 8102 to the degree of water retention stated in the Project Specification. The other components that make up the complete water retention system shall be described in the Project Specification.

B1.9.1 Inspections

The Contractor shall carry out inspections to assess the watertightness of each wall element and each joint between elements. Assessments shall take place whilst construction activities and excavation in front of the wall are being carried out. The watertightness of each wall element and each joint between wall elements shall also be assessed after the ground in front of the wall has been fully excavated to final formation level. The key timings for assessments are presented in the Project Specification. Prior to an inspection to assess the watertightness of a newly revealed portion of the wall, any soil and debris on the front face of the wall shall be removed. The method for removal shall be stated in the Contractor's method statement prior to the commencement of the Works.

The following definitions apply for this Specification:

- Watertightness Assessment Level (WAL) is the lowest level visible on the front face of the wall at the time of a watertightness assessment.
- Damp patch: when touched, a damp patch may leave a slight film of moisture on the hand, but no droplets of water or greater degrees of wetness are left on the hand. On a concrete surface a damp patch is discernible from a darkening of the colour of the concrete.
- Beading of water: beading of water is the state in which individual droplets of water (held by surface tension effects) form on the surface of the wall and adhere to the wall. The water beads do not coalesce with each other. The beads remain stationary on the surface and do not flow.
- Weeping of water: weeping of water is the state in which droplets of water form on the surface of the wall and coalesce with other droplets. The coalesced water does not remain stationary on the wall surface, but instead flows down the wall.

Neither a wall element nor joint between elements shall be considered watertight under this Specification if the following criteria are not met:

1. No weeping of water or greater rates of water ingress or flow is visible between the top of the wall and the WAL. Beading of water is permitted.
2. Damp patches on the front face of the wall are permitted provided that all the following criteria are met:

   (i) the total area of dampness does not exceed 10% of the visible area of the front face;
   (ii) no individual patch of dampness has an area in excess of $4\,m^2$.

Assessment of the watertightness of the wall shall take place only when the groundwater level outside the wall is within the range stated in the Project Specification. In cases where the wall is not protected from weather conditions, assessment of watertightness shall take place only when the following weather conditions apply:

(i) air temperature is above 0°C and below 25°C;
(ii) no direct sunlight falls on any wall surface being assessed for dampness;
(iii) no rainfall has occurred in the 24 hours preceding the assessment.

The groundwater level, air temperature and weather conditions shall be recorded for each assessment.

Assessment of the watertightness of the wall is permitted to take place at the times specified in the Project Specification.

Following completion of the watertightness assessments of the wall and following the satisfactory repair of any leaks encountered, the wall shall be agreed as being 'watertight'.

B1.9.2 Repair

Between the top of the wall and formation level the Contractor shall carry out the repair of any joint in the wall, defect in the wall or section of wall that is not watertight according to the criteria presented above.

The method for repair shall be stated in the Contractor's method statement prior to the commencement of the Works. The Contractor shall include in the programme adequate time for repair works.

Any repair carried out by the Contractor shall be effective for the design life of the wall as stated in the Project Specification.

Leaks arising from water flowing over the top of the wall (including through the joint between the wall and capping beam), through the wall beneath formation level, through the ground beneath formation level, through any joints between slabs and the wall or through any slabs shall not be assessed using the procedures in Clause B1.9.1. Responsibilities for the management and/or repair of these types of leaks shall, where appropriate, be identified in the Project Specification.

B1.10 Construction method

The Contractor shall submit with the tender all relevant details of the method of piling or wall construction, and all relevant details of the plant and monitoring equipment to be used. At least one week prior to commencement of the Works, the Contractor shall submit a detailed method statement to the Engineer.

The method of piling or wall construction shall be such that a continuous monolithic pile or wall element of the specified cross-section is formed to the design depth.

B1.10.1 Obstructions and voids

The Contractor shall submit with the tender all relevant details of the method(s) proposed for dealing with man-made and natural obstructions and voids, when these have been identified in the Project Specification.

B1.10.2 Dimensions of piles and wall elements

The dimensions of a constructed pile or wall element shall not be less than the specified dimensions.

The dimensions of the auger, casing, tool or preformed pile section shall be checked regularly and recorded. A tolerance on these dimensions, e.g. auger diameter, casing diameter, grab length and width, of up to 5% is permissible.

B1.10.3 Concrete casting level tolerances

The heads of concrete piles and embedded walls shall be cast to a level above the specified cut-off so that, after trimming, a sound concrete connection with the pile or wall element can be made.

After casting any empty bore remaining shall be carefully backfilled immediately with inert material.

The completed pile position shall be clearly marked and protected so as not to cause a safety hazard.

B1.10.3.1 Concrete piles or walls constructed in accordance with Sections B3 or B8

For piles cast in dry bores using temporary casing and without the use of a permanent casing, the casting level shall be within the tolerance above the cut-off level shown in Table B1.5, but shall not be above the commencing surface level. No pile shall be cast with its cut-off level below standing water level unless appropriate measures are taken to prevent inflow of water causing segregation of the concrete as temporary casing is extracted.

For piles cast in dry bores where permanent lining tubes or permanent casings are used, or where their cut-off levels are in stable ground below the base of any casing used, the casting level shall be within the tolerance above the cut-off level shown in Table B1.5, but shall not be above the commencing surface level.

Where either support fluid or water is mixed in the ground by the drilling equipment to assist with the installation of temporary casings the casting level shall be in accordance with Table B1.5.

For piles or wall elements cast underwater or support fluid, the pile or wall heads shall be cast to a level within the tolerance above the cut-off level shown in Table B1.5, but shall not be above the commencing surface level. Cut-off levels may be specified below the standing groundwater level, and where this condition applies the water or support fluid level shall not be reduced below the standing groundwater level until the concrete has set.

B1.10.3.2 Concrete piles or walls constructed in accordance with Sections B4 or B5

Concrete shall be cast to the commencing surface level in all cases, unless permanent lining tubes or permanent casings are used. If the level of the fluid concrete subsequently falls below commencing surface level the Contractor shall record the new concrete surface

Table B1.5 Concrete casting tolerance above cut-off levels for specified conditions

Cut-off level below commencing surface H^* (m)	Casting tolerance above cut-off level (m)	Condition
0.15 to any depth	$0.3 + H/10$	Piles cast in dry bore within permanent casing or cut-off level in stable ground below base of casing
0.15–10.00	$0.3 + H/12 + C/8$	Piles cast in dry bore using temporary casing other than above
0.15–10.00	$1.0 + H/12 + C/8$ where C = length of temporary casing below the commencing surface	Piles or walls cast under water or support fluid**

* Beyond $H = 10$ m, the casting tolerance applying to $H = 10$ m shall apply.
** In cases where a pile is cast so that the cut-off level is within a permanent lining tube, or for a wall, the appropriate tolerance is given by deleting the casing term $C/8$.

level, and shall immediately backfill the empty bore with inert material, but not with new concrete.

B1.11 Construction programme

The Contractor shall submit a provisional programme for the execution of the Works at the time of tender and a detailed final programme prior to commencement of the Works. During the Works the Contractor shall inform the Engineer each working day of the intended programme of pile or wall construction for the following working day, and shall give 24 hours notice of intention to work outside normal hours and at weekends, where this is permitted.

B1.12 Records
B1.12.1 Records of the Works

The Contractor shall keep the records listed in Table B1.6 for the installation of each pile or wall element and shall submit one copy of these records to the Engineer within 24 hours after the pile or wall element was completed. The records will form a record of the work. Records may be provided in either hardcopy or electronic format, as agreed with the Engineer.

Any unexpected driving or boring conditions shall be noted in the records.

B1.12.2 Piling and/or Walling Completion Report

Where required in the Project Specification, the Contractor shall submit a 'Completion Report' within 2 weeks of installation of the final pile and/or walling element. The Report shall consist of:

- A summary of the ground and groundwater conditions, together with any variations observed during the works.
- A table for all installed piles/elements with date of installation, actual diameter (or size), actual length, actual base level, assessed penetration into founding strata and actual time between start of boring and completion of concreting.
- For continuous flight auger/displacement auger piles, the concrete volume and the theoretical overbreak shall also be recorded.
- For underream piles, base diameter, level and records of the base condition shall also be recorded.
- For driven piles, the final set and assessed levels for all joints.
- For walls, any comments on the likely water resistance of the wall.
- A summary of integrity and material test records, highlighting the criteria applied and any issues.
- A summary of load test records, highlighting the contractual limits on movements, the basis for selection of the test piles and any issues.
- A summary of the main documents that were used for the Works, together with their revision status; e.g. Specification, drawings, method Statements.
- Full details of all non-conformances together with how they were resolved and their effects on the usability and capacity of all installed piles.
- An assessment of any change in installed capacity for each pile compared to that specified, e.g. due to extra penetration or problems due to non-conformance.
- A list of all piles/elements that were at least partially installed giving details of those abandoned or used for other than as part of the Works (e.g. temporary works).
- A plan showing the final location of all piles/elements.

Table B1.6 Records to be kept (indicated by a tick)

Data	Bearing pile types						Embedded retaining wall types				
	Bored cast-in-place concrete or grout piles	Continuous flight auger and displacement auger concrete or grout piles	Driven pre-cast concrete piles	Driven cast-in-place concrete or grout piles	Driven steel piles	Driven timber piles	Diaphragm walls	Secant pile walls	Contiguous pile walls	King post pile walls	Sheet pile walls
Contract/subcontract reference	✓	✓	✓	✓	✓	✓	✓	✓	✓	✓	✓
Pile/wall element reference number and location	✓	✓	✓	✓	✓	✓	✓	✓	✓	✓	✓
Pile/wall element type	✓	✓	✓	✓	✓	✓	✓	✓	✓	✓	✓
Pile forming equipment including rig type and reference	✓	✓	✓	✓	✓	✓	✓	✓	✓	✓	✓
Nominal cross-sectional dimensions or diameter	✓	✓	✓	✓	✓	✓	✓	✓	✓	✓	✓
Nominal diameter of under-ream/enlarged base	✓			✓							
Length of preformed pile/wall element as appropriate			✓		✓	✓	✓			✓	✓
Date and time of driving, redriving, boring or excavation from start to finish	✓	✓	✓	✓	✓	✓	✓	✓	✓	✓	✓
Date of concreting	✓	✓		✓			✓	✓	✓	✓	
Details of material samples taken	✓	✓	✓	✓	✓	✓	✓	✓	✓	✓	✓
Ground level at pile/wall element position at commencement of installation of pile/wall element (commencing surface level)	✓	✓	✓	✓	✓	✓	✓	✓	✓	✓	✓
Working level on which base machine stands	✓	✓	✓	✓	✓	✓	✓	✓	✓	✓	✓
Depth from ground level or guide wall level, as appropriate, at pile/wall element position to pile/wall element toe	✓	✓	✓	✓	✓	✓	✓	✓	✓	✓	✓
Toe level	✓	✓	✓	✓	✓	✓	✓	✓	✓	✓	✓
Pile/wall element head level as constructed	✓	✓	✓	✓	✓	✓	✓	✓	✓	✓	✓
Pile/wall element cut-off level	✓	✓	✓	✓	✓	✓	✓	✓	✓	✓	✓
Stop end details							✓				

Specification for piling and embedded retaining walls. Thomas Telford, London, 2007

Table B1.6 Records to be kept (indicated by a tick) continued

Record	1	2	3	4	5	6	7	8	9
Length of temporary casing	✓	✓	✓				#	#	#
Length of permanent casing	✓	✓					#	#	#
Groundwater level from direct observation	✓	✓	✓				#	#	#
Type, weight, drop and mechanical condition of hammer and equivalent information for other equipment			✓	✓	✓	✓			
Number and type of packing used and type and conditions of dolly used during driving of the pile			✓	✓	✓	✓			
Set of pile or pile tube in millimetres per 10 blows or number of blows per 25 mm of penetration			✓	✓	✓	✓			
Temporary compression of ground and pile			✓	✓	✓	✓			
Driving resistance taken at 0.25 m intervals			✓	✓	✓	✓			
Length and details of and cover to reinforcement	✓		✓	✓	✓	✓	✓	✓	✓
Level of top of reinforcement cage, as constructed	✓		✓		✓	✓	✓	✓	✓
Concrete or grout mix	✓		✓	✓	✓	✓	✓	✓	✓
Volume of concrete or grout supplied to pile/wall element	✓		✓	✓	✓	✓	✓	✓	✓
Graph of rise of top of concrete or grout level versus volume placed by batch			✓		✓	✓	✓	✓	✓
Depth to average levels of concrete or grout surface before and after withdrawing temporary casing	✓		✓		✓		#	#	#
For raking piles, angle of rake	✓		✓		✓	✓			
Support fluid tests	✓					✓		#	#
Monitoring information and records as referred to in Section B4	✓		✓				#	#	#
Length of offcuts					✓	✓		✓	
Length of pile extensions					✓	✓		✓	✓

Table B1.6 Records to be kept (indicated by a tick) continued

Data	Bearing pile types						Embedded retaining wall types				
	Bored cast-in-place concrete or grout piles	Continuous flight auger and displacement auger concrete or grout piles	Driven pre-cast concrete piles	Driven cast-in-place concrete or grout piles	Driven steel piles	Driven timber piles	Diaphragm walls	Secant pile walls	Contiguous pile walls	King post pile walls	Sheet pile walls
All information regarding obstructions delays and other interruptions to the sequence of work, and unexpected changes in boring, driving or excavation characteristics where appropriate	✓	✓	✓	✓	✓	✓	✓	✓	✓	✓	✓
As constructed positional records vertical and horizontal, if specified in Project Specification	✓	✓	✓	✓	✓	✓	✓	✓	✓	✓	✓
Top and bottom of guidewall level, as appropriate							✓	✓	✓		
Movements of ground and structures and services, if specified in Project Specification	✓	✓	✓	✓	✓	✓	✓	✓	✓	✓	✓
Soil samples taken and in-situ tests carried out during pile/wall element formation or adjacent to pile position, if specified in the Project Specification	✓	✓	✓	✓	✓	✓	✓	✓	✓	✓	✓
Description of ground excavated		✓									
Depth from commencing surface to changes in strata and to standing groundwater and any fluctuations	✓										

means that the type of record will be as required for the piling type being used

Notes.

1. All levels shall be relative to the datum specified in the Project Specification or Ordnance Datum if not specified

2. All times shall be given in 24-hour format

Specification for piling and embedded retaining walls. Thomas Telford, London, 2007

B1.13 Nuisance and damage

B1.13.1 Noise and disturbance

The Contractor shall carry out the work in such a manner and at such times as to minimize noise, vibration and other disturbance in order to comply with current environmental legislation and the noise and vibration limits stated in the Project Specification.

Particular restrictions on permissible working hours shall be stated in the Project Specification.

B1.13.2 Damage to adjacent structures

Permissible damage criteria for adjacent structures or services shall be stated in the Project Specification. If, in the opinion of the Contractor, damage may be caused to other structures or services by the execution of the Works, the Engineer shall be notified immediately. The Contractor shall submit plans for making surveys and monitoring movements or vibration before the commencement of the Works.

The Contractor shall determine the positions of all known services and structures before commencing intrusive work on site.

B1.13.3 Damage to completed piles or wall elements

Piles or wall elements shall not be constructed so close to other piles or wall elements which have been recently formed so that any damage or impairment is caused to the recently formed piles or wall elements.

The Contractor shall submit to the Engineer the planned sequence and timing for driving or boring piles, or installing wall elements, having regard to the avoidance of damage to adjacent works already completed by the Contractor.

The Contractor shall ensure that during the course of the work, displacement or damage which would impair either performance or durability does not occur to completed pile or wall elements. This shall include settlement, heave or lateral displacement of the ground caused by either construction of the Works, the movement of any plant, or the construction activities of other works.

B1.13.4 Temporary support to piles

The Contractor shall ensure that where required, any permanently free-standing displacement piles are temporarily braced or stayed immediately after driving to prevent loosening of the piles in the ground and to ensure that no damage resulting from oscillation, vibration or movement can occur.

The Contractor shall ensure that where required, any temporary or permanently free-standing steel sections or casings are braced or stayed immediately to prevent damage.

B1.14 Driving piles

Clause B1.14 shall only be applicable for piles constructed using the methods described in Sections B2, B5, B6 and B7.

B1.14.1 Driving procedures and redrive checks

The driving procedure shall be such as to avoid damage to the piles. The driving of each pile shall be continuous until the depth or set as required by the design has been reached. In the event of an unavoidable interruption to driving, the pile may be redriven provided it can subsequently be driven to the designed depth and/or resistance or set without damage. A follower shall not be used unless the set is revised where applicable in order to take into account the reduction in the effectiveness of the hammer blow.

Driving records shall be made for every pile. This record shall contain the weight and fall of the hammer or ram and the blows/25 mm of penetration, unless otherwise specified in the Project Specification. The Contractor shall inform the Engineer without delay if an unexpected change in driving characteristics is noted.

B1.14.2 Performance of driving equipment

Where required in the Project Specification, the Contractor shall provide the Engineer with information on the efficiency and energy of the driving equipment, including any followers when used, together with dynamic analysis and evaluation.

B1.14.3 Set

The set and temporary compression shall be measured and recorded for each pile at the completion of driving unless otherwise stated in the Project Specification. When a set is being measured, the following requirements shall be met:

(*a*) The exposed part of the pile shall be in good condition without damage or distortion.

(*b*) The helmet, dolly and any packing shall be in sound condition.

(*c*) The hammer blow shall be in line with the pile axis and the impact surfaces shall be flat and at right angles to the pile and hammer axis.

(*d*) The hammer shall be in good condition, delivering adequate energy per blow, and operating correctly.

(*e*) The temporary compression of the pile shall be recorded if required in the Project Specification.

The set shall be recorded either as the penetration in millimetres per 10 blows or as the number of blows required to produce a penetration of 250 mm.

B1.14.4 Driving sequence and risen piles

Piles shall be driven in a sequence to minimize any detrimental effects of heave and lateral displacement of the ground. The sequence and method of piling, including pre-boring, shall limit vertical and lateral movement so that the final position of each pile is within the specified tolerances. At all times the deflections of each pile from its axis as driven shall not be such as to cause damage or impair durability of the piles or any structures or services.

The maximum permitted uplift of each pile due to any other one pile driven is 3 mm, unless it can be demonstrated by static load pile testing that uplift exceeding this amount does not affect the ability of the pile to meet the requirements of this Specification.

Even if during the installation of preliminary piles uplift is shown to be within the permitted maximum, and the preliminary piles tested meet the requirements of the Specification, checks of uplift on working piles shall be made by the Contractor at least once a week throughout the period of the Works and the results reported to the Engineer.

If preliminary piles are not installed the Contractor shall commence installation of working piles taking measures to reduce or eliminate uplift until it can be established by site measurements that such measures are no longer necessary. Thereafter checks on uplift shall be made by the Contractor at least once a week and the results reported to the Engineer.

If a static load pile test shows that a pile which uplifted more than the maximum permitted amount does not comply with the requirements of the Specification all such preformed piles that have been uplifted shall be redriven. For cast-in-situ concrete piles the Contractor shall submit his proposals in accordance with Clause B1.16.

If records and measurements show that piles have been laterally displaced so as to be outside the permitted tolerance, damaged or are of impaired durability the measures the Contractor plans to adopt to enable the piles to comply with the Specification shall be provided to the Engineer.

B1.14.5 Pre-boring

If pre-boring is specified, the diameter and depth of pre-bore shall be as stipulated in the Project Specification. Other means to ease pile drivability may be used provided the completed piles meet the requirements of the Specification.

B1.15 Supervision and control of the works

The Contractor shall provide a full-time competent supervisor on site to be responsible for the construction aspects of the Works. The Contractor shall state in his tender if this supervisor will not be on site full-time.

This supervisor shall be experienced in the type of construction necessitated by the Contract. A Curriculum Vitae of the supervisor shall be submitted prior to commencement of the Works. The supervisor shall not be removed from the Works without the Engineer being notified in advance with at least one week's notice.

The Contractor shall submit one week prior to commencement of the Works the Quality Plan for the Works. Subsequent revisions, amendments or additions shall be submitted prior to their implementation. Quality Assurance and Quality Control documentation shall be made available on request.

B1.16 Non-conformances

If any part(s) of the Works are discovered or suspected not to conform to this Specification then:

(a) The Contractor shall, within seven days of discovery or receipt of the Engineer's request for further information, notify the Engineer of the proposal(s) and provide calculations or additional information necessary to remedy the non-conformance or substantiate that the work done is of equal value and effectiveness.

(b) The Engineer shall be deemed to have accepted that the Contractor's proposal(s) demonstrate that the work done satisfies the intent of the Specification unless within seven days of receipt of the proposal(s) the Engineer details a request for additional information for this purpose or rejects the proposal(s) with a clear statement describing the reason for the rejection.

B1.17 Trimming and cutting off piles and wall elements

The Contractor shall provide in his tender full details of the method proposed to prepare the heads of piles or wall elements. When trimming or cutting off piles or wall elements down to the specified cut-off level, the Contractor shall ensure that the rest of the pile is not damaged.

For concrete piles or wall elements any laitance or contaminated, cracked or defective concrete shall be removed, and the pile made good in a manner to provide a full and sound section up to cut-off level.

Where debonding of reinforcing bars is specified, the Contractor shall obtain all relevant licenses. Positional tolerances of the debonding material shall be the same as the positional tolerances of the reinforcement.

B1.18 Definitions

In this Specification the terms 'submitted', 'demonstrated', 'notified' and 'required' mean 'submitted to the Engineer', 'demonstrated to the Engineer', 'notified to the Engineer' and 'required by the Engineer' respectively.

Allowable pile capacity: the safe pile capacity taking into account movement requirements. The allowable pile capacity indicates the ability of the pile to meet the specified loading and movement requirements and is therefore required to be not less than the Specified Working Load.

Commencing surface: is the level at which the piling tool or pile first enters the ground. This need not be the same as the working platform level.

Compression pile: a pile which is designed to resist an axial force such as would cause it to penetrate further into the ground.

Contractor: is the principal or main contractor appointed by the Employer to undertake the Works. This could be a specialist piling or walling contractor where they are appointed directly by the Employer. This Specification specifies only the direct contractual responsibilities between the Employer and the Contractor.

Cut-off level: is the level to which the pile is trimmed in order to connect the pile to the structure.

Design Verification Load (DVL): a load which will be substituted for the Specified Working Load for the purpose of a test. This load will be particular to each Preliminary or other test pile and should take into account the maximum Specified Working Load for a pile of the same dimensions and materials, plus allowances for soil induced forces such as downdrag (which may act in reverse under temporary loading conditions), pile spacing, group action and any other particular conditions of the test such as a variation of pile head casting level.

Designer: a qualified engineer or other suitably experienced person who is appointed either by the Employer, the Engineer or the Contractor to carry out design and to issue instructions regarding standards, specification and techniques to be observed in the construction of the Works.

Element: means an individual component utilized in a particular embedded retaining wall system, e.g. diaphragm wall panel, or primary or secondary pile in a secant wall, which can be constructed in isolation.

Embedded retaining wall: means a wall that is constructed in the ground, retains the ground and relies on its embedment for some or all of its stability. It may be (1) a retaining wall of concrete construction with shuttering provided by the surrounding ground (i.e. cast against soil) or (2) steel piles inserted into the ground.

Engineer: a qualified or suitably experienced engineer who is appointed by an Employer to act as their representative on the design, specification and supervision of the Works.

Factor of safety for a pile: the ratio between the pile's ultimate capacity and Specified Working Load. The factor of safety relates only to the failure of the pile–soil interface.

Factor of safety for a wall: the ratio between the restoring forces acting on a wall and the destabilizing forces.

Foundation scheme: means the substructure which may include pile-caps, floor slabs, ground beams and/or basement columns and/or walls.

Geotechnical Design Report: a report, as defined in BS EN 1997-1, that includes the Ground Investigation Report, assumptions, data, calculation methods, calculation results, descriptions, design values, risks and recommendations.

Ground Investigation Report: a report, as defined in BS EN 1997-1, which should comprise all available geotechnical information with an evaluation of this information. This is equivalent to the factual and interpretative Site Investigation reports.

Kentledge: dead load used in a loading test.

King post wall: a wall generally constructed either from steel H-sections driven into the ground or from steel H-sections concreted into bored piles. Pile spacing is normally between 2 m and 3 m. As excavation proceeds horizontal units, which could be made from timber, pre-cast concrete or steel, are inserted between the H-section flanges to support the soil.

Load: a force applied to a pile under test or in service. Equivalent to an action as defined in BS EN 1990, BS EN 1991 and BS EN 1997.

Maintained load test: a loading test in which each increment of load is held constant either for a defined period of time or until the rate of movement (settlement or uplift) falls to a specified value.

Man-made obstruction: a man-made object in the ground which impedes or prevents piling progress.

Natural obstruction: a natural object in the ground which impedes or prevents piling progress.

Pile: all types of bearing pile.

Pile settlement: The axial downward movement at the top of the pile from the position before the commencement of loading. For piles loaded and unloaded through a number of cycles, settlement shall be the cumulative vertical movement.

Pile heave: The axial upward movement at the top of the pile from the position before the commencement of loading. For piles loaded and unloaded through a number of cycles, heave shall be the cumulative vertical movement.

Preliminary pile: an expendable test pile installed before the commencement of the main piling Works or specific part of the Works.

Proof load: a load applied to a selected working pile to confirm that it is suitable for the load at the settlement specified. A proof load should not normally exceed the Design Verification Load plus 50% of the specified Specified Working Load.

Raking pile: a pile installed at an inclination to the vertical.

Reaction system: the arrangement of the kentledge, piles, anchorages or spread foundations that provide a resistance against which the pile is load tested, except for a bi-directional pile load test where the pile itself is used as the reaction.

Representative Action: is defined by BS EN 1990 as the value of the action used for the verification of a limit state.

Resistance: the force developed by a pile in response to a load.

Safe pile capacity: a capacity which takes into account the ultimate pile capacity, the materials from which the pile is made, the required minimum factor of safety, pile spacing, downdrag, and other relevant factors.

Specified Working Load (SWL): the specified load on the head of the pile as stated in the Project Specification. Vertical and horizontal components may be given. In traditional structural terms it is the 'unfactored' rather than 'ultimate' loads. This is similar to the Representative Action defined in BS EN 1990.

Tension pile: a pile which is designed to resist an axial force such as would cause it to be extracted from the ground.

Test pile: any pile to which a test load is, or is to be, applied.

Ultimate pile capacity: the maximum resistance offered by the pile when the strength of the soil is fully mobilized. If the resistance is derived from a test on a Preliminary pile, the settlement rate criteria specified in Clause B15.13.1 shall have been met.

Wall: all types of embedded retaining wall.

Wall Manual: a document, which may form part of the Geotechnical Design Report, comprising, as a minimum, a design summary and a construction sequence drawing.

Working pile: one of the piles forming the foundation of the structure.

Working platform: is the surface, including ramps and access routes, which supports the piling plant and its ancillary equipment. It is also commonly known as the piling platform.

Works: All piling and/or walling activities that are to be undertaken to the standard required by this Specification.

B2 Driven pre-cast concrete piles
B2.1 General

All materials and work shall be in accordance with Sections B1, B2 and B19 of this Specification and BS EN 12699.

B2.2 Project Specification

The following matters are, where appropriate, described in the Project Specification:

(*a*) penetration or depth or toe level
(*b*) driving resistance or dynamic evaluation of set
(*c*) uplift/lateral displacement trials
(*d*) pre-boring or jetting or other means of easing pile driveability
(*e*) detailed requirements for driving records
(*f*) pile shoes (where required)
(*g*) types of pre-stressing tendon
(*h*) marking of piles
(*i*) other particular technical requirements.

B2.3 Materials
B2.3.1 Compliance with standards

In the manufacture of pre-cast reinforced concrete piles, pre-stressed concrete piles and jointed pre-cast reinforced concrete segmental piles, all materials and components shall comply with BS EN 12794 and Section B19. This applies to manufacture of piles in a factory and on site, either indoors or outdoors.

B2.3.2 Pile shoes

Pile shoes shall comply with BS EN 12794.

B2.3.3 Pile head reinforcement

Pile heads shall comply with BS EN 12794.

B2.3.4 Pile quality

A certificate of quality (otherwise known as a evaluation of conformity) shall be provided by the pile manufacturer, in accordance with BS EN 12794.

B2.3.5 Marking of piles

Marking of piles shall comply with BS EN 12794.

B2.4 Construction processes
B2.4.1 Ordering of piles

Piles shall be ordered by the Contractor to meet programme requirements. All piles and production facilities shall be made available for inspection at any time. Piles shall be examined at the time of delivery and any faulty units rejected and replaced.

B2.4.2 Handling, transportation and storage of piles

The method and sequence of lifting, handling, transporting and storing piles shall be such as to avoid shock loading and to ensure that piles are not damaged. Only designed lifting and support points shall be used. During transport and storage, piles shall be supported to avoid damage.

A pile shall be rejected when the width of any transverse crack exceeds 0.3 mm. The measurement shall be made with the pile in its working position.

B2.4.3 Driving piles

Pile installation and procedures shall be in accordance with BS EN 12699 and Clause B1.14.

B2.4.4 Strength of piles

Pile strength at the time of installation shall comply with BS EN 12794.

B2.4.5 Pile installation system

The pile shall be safely supported by the pile installation system.

B2.4.6 Performance of driving equipment

The pile installation system shall be capable of adjustment to provide an axial, non-eccentric application of the installation force.

B2.4.7 Length of piles

The length of each pile driven shall be recorded, and any significant variations in length compared to the design length shall be reported to the Engineer within 24 hours.

B2.4.8 Repair and lengthening of piles
B2.4.8.1 Repair of damaged pile heads

If it is necessary to repair the head of a pile before it has been driven to its final level, the Contractor shall carry out such repair in a way which allows the pile-driving to be completed without further damage. If the driving of a pile has been completed but the level of sound concrete of the pile is below the required cut-off level, the pile shall be made good to the cut-off level, or the pile cap or substructure may be locally deepened, so that the completed foundation will safely withstand the specified working load.

B2.4.8.2 Lengthening of pre-cast reinforced and pre-stressed concrete piles

Any provision for lengthening piles incorporated at the time of manufacture shall be designed by the Contractor to resist all stresses to which it may be subjected.

If no provision for lengthening piles was incorporated at the time of manufacture, any method for lengthening shall be such that the extended pile including any joints is capable of resisting all stresses to which it may be subjected.

Where piles are driven to depths exceeding those expected, leaving insufficient projection for penetration into the following works, the piles shall be extended or replaced so that the completed piles are capable of meeting the requirements of the Specification.

B2.4.8.3 Driving repaired or lengthened piles

Repaired or lengthened piles shall not be driven until cubes for the added concrete have reached the specified characteristic strength of the concrete of the pile.

B2.4.9 Cutting off pile heads

When the driving of a pile has satisfied the Specification, requirements the concrete of the head of the pile shall be cut off to the level specified. Reinforcing bars projecting above this level shall be as specified. Care shall be taken to avoid shattering or otherwise damaging the rest of the pile. Any cracked or defective concrete shall be cut away and the pile repaired to provide a full and sound section to cut-off level.

B2.5 Records

All records shall be in accordance with the requirements of Clause B1.12 and the Project Specification.

B3 Bored cast-in-place piles

B3.1 General

All materials and work shall be in accordance with Sections B1, B3, B18 and B19 of this Specification and BS EN 1536.

B3.2 Project Specification

The following matters are, where appropriate, described in the Project Specification:

(*a*) support fluid
(*b*) base or shaft grouting
(*c*) pile shaft and base inspection by CCTV and or sampling/probing (see Clause B3.5.8)
(*d*) details of permanent casings
(*e*) inspection of underreams
(*f*) other particular technical requirements.

B3.3 Materials
B3.3.1 Support fluid

Where support fluid is used to maintain the stability of the excavation, it shall be in accordance with Section B18 of this Specification.

B3.3.2 Casings

Temporary casings shall be of quality of material, length and thickness adequate for the purpose of preventing water and unstable soil from entering the pile excavations. A short length of temporary casing shall be provided for all piles to provide an up-stand of at least 1 m above surrounding ground level for safety and to prevent contamination of the concrete in the bore.

Temporary casings shall maintain the excavations to their full dimensions and ensure that piles are completed to their specifed cross-sectional dimensions.

Temporary casings shall be free from significant distortion. They shall be of uniform cross-section throughout each continuous length. During concreting they shall be free from internal projections of any kind which might adversely affect the proper formation of piles.

If a temporary casing is to be used to core through obstructions (or adjacent piles for secant walls) then the Contractor shall submit details of any cutting head to be used on the leading edge of the casing which protrudes outside of the edge of the casing at the time of tender.

Permanent casings shall be in accordance with Clause B16.4.

B3.4 Construction tolerances
B3.4.1 Steel reinforcement

Reinforcement shall be maintained in its correct position during concreting of the pile, to allow a vertical tolerance of $+150/-50$ mm (i.e. a maximum of 150 mm high) on the level of the reinforcement projecting above the final cut-off level. Where reinforcement is made up into cages, they shall be sufficiently rigid to enable them to be handled, placed and concreted without damage. Details of the procedures shall be submitted to the Engineer one week prior to commencement of the Works. Unless otherwise specified, the Contractor shall extend the reinforcement to the base of the pile or to 3 m below the bottom of the temporary casing to ensure that the construction tolerances are achieved.

B3.5 Construction processes
B3.5.1 Placing of casings

The use of a vibrator to insert and withdraw temporary casings is subject to compliance with Clause B1.13 and to the method not causing disturbance of the ground which would adversely affect the construction or the capacity of piles, or damage to adjacent structures.

Where piles are bored under water or support fluid in an unlined state, the insertion of a full-length loosely fitting casing to the bottom of the bore prior to placing concrete will not be permitted.

Permanent casings shall be as specified in the Project Specification.

Where the use of a permanent casing is specified, the Contractor shall submit details of the type of casing and the method of installation in the method statement at the time of tender, including measures to backfill any annulus around the permanent casing.

B3.5.2 Placing concrete
B3.5.2.1 General

The concrete shall be placed without such interruption as would allow the previously placed batch to have achieved a stiffness which prevents proper amalgamation of the two concrete batches.

The Contractor shall take all precautions in the design of the mix and placing of the concrete to avoid arching of the concrete in a temporary casing. No soil, liquid or other foreign matter shall be permitted to contaminate the concrete.

B3.5.2.2 Consistence of concrete

The consistence of the concrete shall be in accordance with Section B19, and the method of placing concrete shall be such that a continuous monolithic concrete shaft of the specified cross-section is formed.

Concrete shall be self-compacting. Internal vibrators shall not be used to compact concrete.

B3.5.2.3 Placing concrete in dry borings

Measures shall be taken to ensure that the structural strength of the concrete placed in the piles is not impaired through grout loss, segregation or bleeding.

The method of placing shall be such as to ensure that the concrete in its final position is dense and homogeneous. Concrete shall be introduced into the pile via a hopper and suitable length of rigid delivery tube to ensure that the concrete falls vertically and centrally down the shaft. The tube shall be of sufficient length to ensure that concrete falls freely no more than 10 m through any reinforcement cage.

B3.5.2.4 Placing concrete under water or support fluid

Measures shall be taken to ensure that the structural strength of the concrete placed in the piles is not impaired through grout loss, segregation or bleeding. The method of placing shall be such as to ensure that the concrete in its final position is dense and homogeneous.

Before placing concrete, measures shall be taken in accordance with Clauses B3.5.7 and B3.5.8 to ensure there is no accumulation of drilling spoil or other material at the base of the boring, and the Contractor shall ensure that heavily contaminated support fluid, which could impair the free flow of concrete from the tremie pipe, has not accumulated in the bottom of the bore.

Concrete to be placed under water or support fluid shall be placed by a full depth tremie in one continuous operation and shall not be discharged freely into the water or support fluid. Pumping of concrete may be used where appropriate. The bottom end of the tremie shall be square to the axis of the tremie and its circumference shall be continuous. The tremie shall be fully charged with concrete before it is lifted off the base of the pile.

The depths to the surface of the concrete shall be measured and the length of the tremie in use recorded at regular intervals corresponding to the placing of each batch of concrete. The depths measured and volumes placed shall be plotted immediately on a graph and compared with the theoretical relationship of depth against volume.

The hopper and pipe of the tremie shall be clean and watertight throughout. The tremie pipe shall extend to the base of the bore and a sliding plug or barrier shall be placed in the pipe to prevent direct contact between the first charge of concrete in the tremie and the water or support fluid. The pipe shall at all times penetrate the concrete which has previously been placed with a minimum embedment of 3 m and maximum of 6 m and shall not be completely withdrawn from the concrete until completion of concreting. A sufficient quantity of concrete shall be maintained within the tremie pipe to ensure that the pressure from it exceeds that from the water or support fluid. The internal diameter of the tremie pipe shall be of sufficient size to ensure the easy flow of concrete. It shall be so designed that external projections are minimized, allowing the tremie to pass within reinforcing cages without causing damage.

B3.5.3 Stability of pile bore

Where boring takes place through unstable and/or water-bearing strata, the process of excavation and the support fluid and depth of temporary casing where employed shall be such that soil from outside the area of the pile bore is not drawn into the pile section and cavities are not created outside the temporary casing as it is advanced.

Where a support fluid is used for maintaining the stability of a bore, a temporary casing shall be used in conjunction with the support fluid so as to ensure stability of the strata near ground level until concrete has been placed. During construction, the level of support fluid in the pile excavation shall be maintained, whether the bore is cased or not, so that it is not less than 2 m above the level of external standing groundwater at all times, unless stated otherwise in the Project Specification.

In the event of a loss of support fluid from a pile excavation, the Contractor shall notify the Engineer within 24 hours of the intended action before continuing.

B3.5.4 Pumping from pile bores

Pumping from pile bores shall not be permitted unless the bore has been sealed against further water entry by casing or unless the soil is stable and will allow pumping to take place without ground disturbance below or around the pile.

B3.5.5 Continuity of construction

The pile shall be bored and the concrete shall be placed without such delay as would lead to impairment of the performance of the pile, or affect its compliance with the design assumptions.

The time period during which each pile is excavated and the concrete is placed shall not exceed 12 hours, unless permitted in the Project Specification. The time period shall start when excavation below the temporary casing commences (except for casings which leave an annulus between the casing and the soil). Where casings are used which leave an annulus around their perimeter, the time period shall start when the leading edge of the casing passes into a stratum which contributes to the load-carrying capacity of the pile.

Where the construction sequence is such that the time period of 12 hours will be exceeded even if no delays are taken into account, a realistic time period, during which the pile is excavated and concrete placed, shall be stated in the Contractor's method statement.

The Contractor shall advise the Engineer on the likely effect of any annulus formed during casing installation, or any extension of the pile construction period on the performance and capacity of the pile. The Engineer (Option 1) or Contractor (Option 2), see Clause B1.4, shall consider this, and modify the pile design if appropriate.

B3.5.6 Underreams

A mechanically formed enlarged base shall be no smaller than the dimensions specified and shall be concentric with the pile shaft to within a tolerance of 10% of the shaft diameter. The sloping surface of the frustum forming the enlargement shall make an angle to the axis of the pile of not more than 35°. At the specified diameter of the underream at the perimeter of the base there shall be a minimum height of 150 mm.

A tool shall be used which has been previously proven capable of producing underream bases that meet this Specification. The Contractor shall measure and record the pile depth after every penetration of the underreaming tool in order to confirm that the final use of the tool is the deepest cut of the base. The measuring method shall be accurate to 15 mm.

A device to confirm the diameter of the underream shall be used, accurate to the tolerances given in Clause B1.10.2.

B3.5.7 Cleanliness of pile bases
B3.5.7.1 General

On completion of boring, loose, disturbed or softened soil shall be removed from the bore using appropriate methods, which shall be designed to clean while at the same time minimize ground disturbance below the pile base. Where used, support fluid shall be maintained at such levels throughout and following the cleaning operation that stability of the bore is preserved.

B3.5.7.2 Underreams

It is a design requirement that the founding material for the underream base is intact, undisturbed soil, and that the shelf formed is substantially free of remould material and debris. All underream piles shall therefore be inspected in detail to verify the design requirement. A remote inspection method using a testing device shall be employed which can clearly demonstrate that the pile base has achieved the requirements of the Specification. Regardless of the method of inspection it shall be capable of demonstrating that at least 90% of the plan area of the base is free of debris and/or remoulded clay.

B3.5.8 Inspection
B3.5.8.1 General

Each pile bore which does not contain support fluid shall be inspected from the ground surface, using a light which is sufficiently powerful to allow the base to be viewed clearly, prior to concrete being placed, to ensure the base is adequately clean and that the shaft is stable and within the specified tolerances. Adequate means of lighting, measuring tapes and a means of measuring verticality shall be used. The pile shaft may alternatively be inspected by CCTV.

Manned inspections of bores are not permitted.

B3.5.8.2 Underreams

Manned inspections of underreams are not permitted. The inspection of underreams shall be carried out remotely using CCTV.

The CCTV inspection shall be subject to the following requirements:

- A high-resolution colour camera shall be used (minimum pixel matrix: 512×582; sensitivity of lens: 1 lux at F1.2; vertical resolution: 310 lines; horizontal resolution: 380 lines).
- Powerful lighting shall be provided.
- The camera shall have pan and tilt capability so that the whole of the base can be viewed as well as the location of the sampling tool.
- The viewing screen shall be a minimum of 250 mm by 250 mm, and compatible with the camera such that clear, well-defined images are produced.
- A recording of the inspection shall be made.
- Lowering and raising of the camera shall be controlled.

A testing device (sampler or penetrometer) capable of confirming the absence of remoulded material on the underream shelf at locations selected by the Engineer shall be employed. At least four tests shall be undertaken for each base. The testing device shall be able to detect the presence of remoulded material.

If the testing device is a sampling device, it shall be subject to the following requirements:

- The sampling tool shall recover samples from the shelf of the underream.
- It shall be possible to guide the sampler using CCTV.
- The sample shall be of sufficient size to give good representation of the area, with a minimum sample diameter of 100 mm.
- Sampling depth shall be at least 200 mm, and the sampler shall not compress the top of the sample.
- The sampler shall be designed to retain any lumps of clay which are present on the surface of the underream shelf at the sampling location and these shall not be discarded.
- The sample shall be extruded on site for examination and testing by pocket penetrometer or hand vane prior to concreting.
- The Contractor shall provide equipment to enable a 'clean' cut of the sample to be made to allow inspection of the material fabric.

If the testing device is a penetrometer or probe, it shall be subject to the following requirements:

- The penetrometer/probe tool shall test the shelf of the underream.
- It shall be possible to guide the penetrometer/probe using CCTV.
- The penetrometer/probe shall be capable of demonstrating the difference between intact and remoulded soil.
- The penetrometer/probe shall be capable of demonstrating the variation of penetration resistance with depth of penetration.

The Contractor shall demonstrate previous experience with the equipment to be used. Case history data showing the successful operation of the testing device is required, together with evidence of its ability to distinguish between an acceptable and an unacceptable base.

In addition to the tests and inspections carried out, if there is any cause for concern or a delay occurs of more than 2 hours between inspection and start of concreting, the CCTV camera shall be lowered to the base for a final inspection before concreting. This is to confirm there has been no substantial collapse over this time. Again the images shall be recorded.

If the inspection is not satisfactory then the underream tool shall be re-introduced into the pile bore, the base cleaned further, and the inspection process repeated.

B3.5.9 Extraction of casing
B3.5.9.1 Workability of concrete

Temporary casings shall be extracted while the concrete within them remains sufficiently workable to ensure that the concrete is not lifted. During extraction the motion of the casing shall be maintained in an axial direction relative to the pile.

B3.5.9.2 Concrete level

When the casing is being extracted, a sufficient quantity of concrete shall be maintained within it to ensure that pressure from external water, support fluid or soil is exceeded and that the pile is neither reduced in section nor contaminated.

The concrete level within a temporary casing in a dry bore shall be topped up, where necessary, during the course of casing extraction so that the base of the casing is always below the concrete surface until the casting of the pile has been completed.

Adequate precautions shall be taken in all cases where excess heads of water or support fluid could occur as the casing is withdrawn because of the displacement of water or fluid by the concrete as it flows into its final position against the walls of the pile bore.

Where two or more discontinuous lengths of casing (double casing) are used in the construction the Contractor's method of working shall produce piles to their specified cross-sections.

The depth to the average levels of the concrete surface of the pile shall be measured before and after each temporary casing is removed. These measurements shall be recorded on the pile record.

B3.5.10 Protection to pile bores

At all times when the pile head is unattended, the bore shall be clearly marked and fenced off.

The completed pile position shall be clearly marked and protected so as not to cause a safety hazard.

B3.5.11 Shaft or base grouting
B3.5.11.1 Grouting of piles

Where bases or sides of piles are to be pressure grouted, if required by the Project Specification, the Contractor shall construct the piles with grout tubes and any other necessary equipment pre-installed so that piles may subsequently be grouted.

B3.5.11.2 Method of grouting

The method of grouting shall be such that the completed pile meets the requirements of the Project Specification for load-settlement behaviour and that during grouting pile uplift is within the limits specified. The Contractor shall submit full descriptions of the equipment, materials and methods that are to be used. The method statement shall comprise at least the following information:

(*a*) Details of the specialist contractor for grouting (if applicable), names of key personnel and their Curricula Vitae and previous experience on similar types of work.
(*b*) Details of grout pump, mixer, agitator and any other equipment used for mixing and injection of grout.
(*c*) Full details of grout to be used, including additives.
(*d*) Method of quality control on grout, including details of number of cubes taken and checks on density, flow and bleed of the grout.
(*e*) Method of measuring grout take, which should be automatic and include a physical method of checking grout take at the end of injection of each circuit.
(*f*) Method of measuring grout pressures which should include a continuous record; calibration certificates for pressure gauges.
(*g*) Typical record sheet for grouting, which shall include records of grout take, grout pressure, residual pressure, times of grouting and pile uplift for each grouting circuit; typical continuous records of grout pressure and pile uplift shall also be included.
(*h*) Target minimum, maximum and residual grout pressures for each grout injection.
(*i*) Method of grout injection, including full details of any packers; target grout volumes for each injection.
(*j*) Method of measuring friction losses in the tubes/packers.

B3.5.11.3 Grout system

The grouting tubes shall be tested to determine any grout leakage in joints under pressure prior to installation into the piles. The grout tubes shall be capable of withstanding the pressures to which they will be subjected.

Robust threaded caps shall be provided to protect the top of the grouting tubes during concreting and afterwards.

The grouting tubes shall be flushed with water after each grouting operation. If the target grout pressures are not achieved, or the specified uplift is not achieved, then the pile shall be regrouted within 24 hours.

Prior to commencing grouting, the Contractor shall demonstrate the grouting system's compliance. The Contractor shall calibrate the delivery system by measuring the stiffness of the hoses, couplings, and grout tubes etc. This shall be taken into account when assessing the volume of grout injected.

During grouting, the Contractor shall provide a competent and suitably trained person to monitor grout pressures and grout takes.

B3.5.11.4 Pile uplift

During base grouting, the pile uplift shall be not less than 0.2 mm and shall not exceed 2.0 mm.

The Contractor shall provide a competent and suitably trained person to monitor the uplift of the piles during grouting.

Uplift of the pile head and base shall be measured during base grouting.

The uplift of the pile head shall be measured by two independent instruments. These shall be a displacement transducer and a dial gauge measuring displacements relative to a reference beam.

The level of the reference beam shall be checked by precise level, both before and after the base grouting operation. The supports of the reference beam shall be at least 2 m from the edge of the pile.

Uplift of the pile base shall also be measured at an elevation 0.5 m above the pile base using two retrievable extensometers installed within tubes cast into each pile. The proposed extensometer system shall be submitted by the Contractor with the tender for approval by the Engineer. Extensometers cast into the pile shall be in accordance with Clause B17.3.1.2.

Details of the measurement system shall be submitted by the Contractor with the tender and shall be subject to acceptance by the Engineer.

B3.5.11.5 Grout testing

Close control of the mixing of the grout shall be carried out. The Contractor shall provide and maintain on site all test facilities required to test and control the grout mixes, in accordance with the Contractor's Method Statement.

B3.6 Records
B3.6.1 *Records of piling*

All records shall be in accordance with Clause B1.12 and the Project Specification.

B3.6.2 *Grouting records*

The Contractor shall provide to the Engineer one copy of all grouting records for each pile within 24 hours of the completion of grouting that pile.

The records shall comprise the following information:

(*a*) Pile number.
(*b*) Date.
(*c*) Leakage test on grout pipes.
(*d*) Grout mix.
(*e*) Continuous records of grout pressure and pile uplift to the same timescale. The records shall be annotated at regular intervals with physical measurements from pressure gauges and dial gauges including times of such measurements. Circuit number and pile number are also to be included.

(*f*) For each circuit the peak and residual grout pressures including times and sequence of grouting.

(*g*) The total pile uplift and the uplift after each circuit is grouted.

(*h*) For each circuit the physical and automatic measurements for grout volume injected and derived injected volumes; the total grout volume.

(*i*) All tests made on grout.

Specification for piling and embedded retaining walls. Thomas Telford, London, 2007

B4 Piles constructed using continuous flight augers or displacement augers

B4.1 General

All materials and work shall be in accordance with Sections B1, B4 and B19 of this Specification and BS EN 1536.

This section applies to piles concreted through a hollow auger stem. The following types of piles are covered:

(i) Continuous flight auger piles (CFA) which employ a continuous flight auger for both advancing the bore and providing support to the surrounding soil.

(ii) Displacement auger piles (DA) which employ a displacement tool with following tube, which is screwed into the ground displacing the surrounding soil.

B4.2 Project Specification

The following matters are, where appropriate, described in the Project Specification:

(a) permitted pile types
(b) whether splitting of augers shall be permitted
(c) whether concrete supply is to be controlled by pressure or volume (see Clause B4.4.5.3)
(d) detailed requirements for monitoring records
(e) other particular technical requirements.

B4.3 Materials
B4.3.1 Steel reinforcement

The reinforcement shall be fabricated as cages, bundles of bars, or steel sections fixed securely to permit it to be placed in the correct position and to the depth specified into the fluid pile concrete.

B4.4 Construction processes

The rig operator shall be competent and experienced in the construction of the specified type of piles. A supervisor shall be in attendance during pile construction. Details of relevant experience of the rig operator and the supervisor shall be submitted prior to work commencing.

B4.4.1 Boring

The piles shall be constructed using equipment capable of penetrating the ground to the design depth. Piles shall be constructed without drawing surrounding materials laterally into the pile bore.

The pile shall be constructed in a manner to minimize the occurrence of flighting or heave of the ground such that no detrimental effects occur. If these effects are observed then the fact shall be reported to the Engineer within 24 hours and recorded in the pile record.

If during augering it is necessary to raise the auger and subsequently to re-auger, the required depth shall be increased to at least 0.5 m below the depth previously reached if this is practical, and the fact shall be recorded on the pile record. The Engineer shall be notified of the occurrence within 24 hours.

B4.4.2 Sealing the base of the auger

The base of the auger stem shall be fitted with a suitable means of sealing against ingress of water and soil during boring.

B4.4.3 Removal of augers from the ground

The auger shall not be extracted from the ground during the construction of a pile in such a way that at any depth an open or unsupported bore or inflow of water into the pile section prior to concrete placement would result.

During concreting the Contractor shall ensure that the base of the auger is always embedded within fluid concrete. Concrete shall overflow at the ground surface before the auger is removed from the ground.

As the auger is withdrawn from the ground it shall be cleaned of all rising spoil in a safe manner and before the spoil reaches a height at which its falling from the auger represents a hazard.

B4.4.4 Depth of piles

Any failure of a pile to reach the design depth, shall be reported to the Engineer with a full statement of the reasons within 24 hours.

B4.4.5 Placing concrete
B4.4.5.1 General

The concrete shall be suitable for pumping. The workability and method of placing shall be such that a continuous monolithic concrete shaft of the specified cross-section is formed.

Concrete shall be supplied to the pile through a suitable concrete pump, tubing and the hollow auger stem. All pipe fitments and connections shall be so constructed that the system does not leak during the injection process.

Concrete shall be self-compacting. Internal vibrators shall not be used to compact concrete. Concrete shall be cast to the commencing surface level unless a cased CFA system is used. If a cased CFA system is used, concrete shall be cast to the tolerances set out in Table B1.5.

B4.4.5.2 Commencement of concrete supply to each pile

Prior to the commencement of concreting and the removal of augers from the ground, the hollow auger delivery pipe and tubing shall be pre-charged with concrete. At the beginning of concrete placement the sealing device shall be removed.

For CFA piles, the auger shall be lifted the minimum distance necessary to initiate the flow of concrete, and no more than 100 mm, such that water inflow and soil movement and collapse at the base of the auger are minimized. Once concrete flow has been initiated, if the auger has been lifted, the auger shall be returned to the maximum depth of the bore, drilling through the initial concrete to ensure a competent base to the pile.

For DA piles the Contractor shall submit at the time of tender details of the procedure to be followed to initiate the delivery of concrete.

The technique and equipment used to initiate and maintain the concrete flow shall be such that a pile of the full specified pile section is obtained from the maximum depth of boring to the commencing surface level.

B4.4.5.3 Rate of supply of concrete

The concrete shall be supplied to the pile at a sufficient rate during auger withdrawal to ensure that a continuous shaft of at least the specified pile cross-section is formed, free from debris or segregated concrete.

If rotation of the auger occurs during extraction, it shall be in the direction that would cause the auger to penetrate into the concrete. If for a DA pile it is necessary to reverse the auger rotation during extraction, this shall be clearly stated by the Contractor in the method statement, along with the reasons why this is necessary, and the measures to be taken to prevent the introduction of defects into the completed pile.

The method for the control of concreting that will be used for the construction of the Works shall be stated in the Project Specification and shall be one of the following two methods:

Method 1 — Control of concreting by measuring concrete pressure
A positive pressure shall be maintained at the point of measurement at all times during concreting. The magnitude of the positive pressure shall always be greater than the ground and groundwater pressure at the base of the auger.

Method 2 — Control of concreting by measuring volume above ground
An oversupply of concrete shall be maintained at the point of

measurement of volumetric flow of concrete at all times during concreting. The volume of the concrete supplied shall be sufficient to ensure a continuous flow of concrete in excess of the theoretical pile volume including an additional allowance for overbore and over-break.

Preliminary piles shall be constructed using the same method of concrete control as is to be used for the working piles.

On completion of concreting, the Contractor shall record on the pile record if the concrete level drops in the bore upon removal of the auger from the ground. If the level of the fluid concrete subsequently falls below casting level the Contractor shall record this level, and not fill the bore with fresh concrete.

B4.4.5.4 Interruption in concrete supply

If the concreting of any pile cannot be completed continuously in the normal manner, then the pile shall be re-augered to a safe level below the position of interruption of supply before concrete has achieved initial set and before any further concrete is injected. Records should be provided for both the original and re-bored pile elements. The method statement shall set out the site specific procedure for determining what the safe level is.

B4.4.6 Splitting of augers

Lengths of auger shall not be split during auger extraction other than where permitted in the Project Specification.

If a methodology is used that requires the splitting of augers, total concreting time per pile shall not exceed 1 hour.

When constructing a pile using the split auger method, the concrete level shall not be permitted to fall lower than the auger joint such that fresh concrete shall overspill the joint as the auger lengths are split. The auger joint shall be at least 1 m above commencing surface level.

B4.4.7 Placing of reinforcement

All reinforcement shall be placed with the minimum delay after the completion of the concreting operation.

Reinforcement shall be placed and maintained in position to provide the specified projection of reinforcement above the final cut-off level. A vertical tolerance of $+150/-50\,\mathrm{mm}$ (i.e. a maximum of 150 mm high) on the level of reinforcement projecting above the final cut-off level shall be met.

DA piles shall be reinforced to resist any heave forces that may occur as a result of installation.

Where reinforcement is made up into cages, they shall be sufficiently rigid to enable them to be handled, placed and concreted without damage.

B4.4.8 Continuity of construction

The pile shall be bored and the concrete shall be placed without such delay as would lead to impairment of the performance of the pile, or affect its compliance with the design assumptions.

B4.4.9 Monitoring system for pile construction
B4.4.9.1 Automated monitoring system

An automated system shall be provided for monitoring the construction of the piles, unless a method using splitting of the auger has been permitted in which case the monitoring system need not be fully automated, but shall measure all of the required parameters. The automated monitoring system shall, as a minimum, provide the operator with continuous real-time information on the depth of the base of the auger, volumetric flow of concrete and concrete pressure.

The monitoring equipment shall monitor continuously with depth the following parameters as a minimum:

During boring
  (i)   depth of the base of the auger
  (ii) time
  (iii) auger revolutions
  (iv) piling rig torque

During concreting
  (i)   depth of the base of the auger
  (ii) time
  (iii) auger revolutions
  (iv) concrete pressure
  (v)   concrete volume

Based on these measured parameters, the Contractor shall ensure that the monitoring system presents the measurements in the following formats on the pile records:

During boring
  (i)   auger penetration rate versus depth
  (ii) auger revolutions per metre penetration versus depth
  (iii) applied rig torque

During concreting
  (i)   rate of extraction of the auger versus depth
  (ii) auger revolutions per metre penetration versus depth
  (iii) relative concrete pressure versus depth
  (iv) rate of supply of concrete versus depth

The Contractor shall state in the method statement whether the measurements from the monitoring system and the pile records are to be provided to the Engineer in hardcopy or electronic format.

In the case of DA piles, the Contractor shall provide in the method statement full details of any additional monitoring that will need to be carried out to ensure that a pile of the specified pile section is formed.

The automated monitoring system shall be fully operational at the start of every pile. If a breakdown of the monitoring system occurs, no further piles are to be constructed until it has been repaired.

To facilitate independent checking of pile construction the following shall be provided in addition to the automated monitoring system:

  (i)   The mast of the piling rig shall have paint marks at 0.25 m intervals over its full height and the metre increments shall be marked numerically.
  (ii) A method for recording the volume or rate of concrete supply.

B4.4.9.2 Manual monitoring system

A manual monitoring system shall only be employed if required for piles constructed using split-auger techniques, or for the completion of a pile where the automated instrumentation system has failed during pile construction.

The Contractor's method statement shall include the procedure to be followed and identify a suitably qualified person responsible for manual monitoring. In the case of automated system failure, the procedure shall additionally record the depth at which failure occurred, the time for auger extraction during concreting, and the total volume of concrete delivered.

Any pile which was subject to an instrumentation failure during construction shall be integrity tested in accordance with Section B.13.

B4.4.9.3 Target boring and concreting parameters

Prior to the commencement of piling, the Contractor shall inform the Engineer of values of the following parameters:

(i)   Target minimum and maximum values of CFA auger revolution per metre penetration during boring of piles.

(ii) Target minimum and maximum percentage concrete oversupply during concreting of piles if an oversupply method is to be used.

B4.4.9.4 Calibration

Equipment used for monitoring shall be calibrated at the start of the Works and calibration certificates issued to the Engineer prior to commencement of the piling. After the commencement of the Works, the monitoring equipment shall be calibrated at the frequency specified below or at any time when there is reason to suspect malfunction.

Depth shall be calibrated at the start of the works and once a week. The tolerance on full auger string length shall be ±0.1 m.

Concrete volume shall be calibrated at the start of the Works and once a week by passing a known volume of concrete through the system. The tolerance on volume is ±5%. If the concrete discharge pumps are changed or replaced on site, and/or the length of concrete delivery tubing is altered, and/or the concrete mix is changed, then this calibration shall be repeated immediately.

The pressure transducer shall be calibrated at an interval of no more than six months. The calibration certificate shall be available for inspection by the Engineer at the commencement of the Works. The tolerance on pressure shall be stated in bars.

B4.5 Records
B4.5.1 Records of piling

The Contractor shall provide such records as required by Clause B1.12, B4.4 and the Project Specification.

In addition to these requirements, the following information shall be recorded during the construction of each pile and reported:

(i)   Final depth of the bore.

(ii)   The target percentage oversupply for that pile (if an oversupply method is to be used to control concrete placement).

(iii) Time for each 0.5 m increment of auger extraction during concreting.

(iv)   The volume of concrete pumped for each 0.5 m increment of auger extraction during concreting.

(v)   The total volume of concrete for completion of the pile.

(vi)   The time at the start and end of boring and concreting, insertion of the reinforcement cage, and any time intervals for delays or stoppages.

(vii) Concrete pressure at point of measurement.

(viii) The direction of rotation of the auger during concreting.

(ix)   The volume of concrete supplied divided by the theoretical pile volume.

(x)   Whether any pull down force was used during pile construction.

(xi)   Any delays/stoppages during drilling/concreting.

(xii) Whether concrete is observed to be flowing under pressure out of the top of the bore prior to the tip of the boring tool emerging.

The raw monitoring data shall be made available to the Engineer immediately if requested, and full monitoring records within 48 hours of completion of each pile. Any anomalous results shall be highlighted by the Contractor.

B5 Driven cast-in-place piles
B5.1 General

All materials and work shall be in accordance with Sections B1, B5 and B19 of this Specification and BS EN 12699.

B5.2 Project Specification

The following matters are, where appropriate, described in the Project Specification:

(a) penetration or depth or toe level
(b) driving resistance or dynamic evaluation or set
(c) uplift/lateral displacement trials
(d) pre-boring or jetting or other means of easing pile driving
(e) detailed requirements for driving records
(f) types and quality of pile shoes
(g) sampling and testing of pile materials
(h) other particular technical requirements.

B5.3 Materials
B5.3.1 Pile shoes

Pile shoes shall comply with BS EN 12699.

B5.3.2 Concrete for use with base driving or for an enlarged base

A plug consisting of semi-dry concrete minimum grade 20 with a water/cement ratio not exceeding 0.25 shall be used when base driving or for the formation of an enlarged base.

B5.3.3 Reinforcement

Reinforcement shall be provided to the full depth of the pile.

B5.4 Construction processes
B5.4.1 Temporary casings

Temporary casings shall be free from significant distortion. They shall be of uniform cross-section throughout each continuous length and shall be of sufficient strength to withstand driving and ground forces without distortion. They shall be free from internal projections which might prevent the proper formation of piles.

B5.4.2 Length of piles

The length of each pile driven shall be recorded, and any significant variations in length compared to the design length shall be reported to the Engineer within 24 hours.

B.5.4.3 Driving piles

Pile installation and procedures shall be in accordance with the requirements of Clause B1.14.

B5.4.4 Base driving

When base driving the concrete plug shall have compacted height of not less than 2.5 times the diameter of the pile. Fresh concrete shall be added to ensure that this height of driving plug is maintained in the casing throughout the period of driving.

B5.4.5 Enlarged pile bases

Where the Contractor wishes to form a pile with an enlarged base or where such a base is specified in the Project Specification, details of his method of forming the base and the materials to be used shall be submitted with the Contractor's method statement.

B5.4.6 Repair of damaged pile heads and extension of piles

When repairing or extending the head of a pile, the head shall be cut off square at sound concrete, and all loose particles shall be removed by wire-brushing followed by washing with water.

If the level of sound concrete is below the cut-off level, the pile shall be extended to the cut-off level with concrete of a grade not inferior to that to the concrete of the pile.

B5.4.7 Inspection and remedial work

Prior to placing concrete in a casing, the Contractor shall check that the casing is undamaged and free from water or other foreign matter. In the event of water or foreign matter having entered the casing, the casing shall be withdrawn, repaired if necessary, and re-driven or other action taken to continue the construction of the pile to meet the requirements of the Project Specification.

B5.4.8 Placing concrete

The workability and method of placing the concrete shall be such that a continuous monolithic concrete shaft of the specified cross-section is formed. Concrete shall be transported from the mixer to the pile position in such a manner that segregation of the mix does not occur.

Fresh concrete shall only be poured into concrete which has retained its full consistence.

The contractor shall take all precautions in the mix design and placing of the concrete to avoid arching of the concrete in the casing. No spoil, liquid or other foreign matter shall be permitted to contaminate the concrete.

Internal vibrators shall not be used to compact concrete cast-in-place. Measures shall be taken to ensure that the structural strength of the placed concrete is not impaired through grout loss, segregation or bleeding.

B5.4.9 Extraction of casing

Temporary casings shall be extracted while the concrete within them remains sufficiently workable to ensure that the concrete is not lifted. Where a semi-dry concrete mix is used, the Contractor shall ensure that the concrete is not lifted during extraction of the casing.

When the casing is being extracted, a sufficient quantity of concrete shall be maintained within it to ensure that pressure from external water or soil is exceeded and that the pile is neither reduced in section nor contaminated.

Concrete shall be topped up as necessary while the casing is extracted until the required head of concrete to complete the pile in a sound and proper manner has been provided.

B5.4.10 Vibrating extractors

Vibrating casing extractors shall be permitted subject to compliance with Clause B1.13.

B5.4.11 Concrete casting tolerances

For piles constructed without the use of a rigid permanent sleeve, concrete shall be cast to the commencing surface level.

Where piles are constructed with the use of a rigid permanent sleeve, pile heads shall be cast to a level above the specified cut-off such that, after trimming, a sound concrete connection with the pile can be made. In this case the tolerance of casting above the cut-off level shall be as required by Clause B1.10.3.

B5.4.12 Cutting off pile heads

When cutting off and trimming piles to the specified cut-off level, the Contractor shall take care to avoid shattering or otherwise damaging the rest of the pile. Any cracked or defective concrete shall be cut away and the pile repaired to provide a full and sound section to the cut-off level.

B5.5 Records
B5.5.1 Record of piling

All records shall be in accordance with the requirements of Clause B1.12 and the Project Specification.

Specification for piling and embedded retaining walls. Thomas Telford, London, 2007

B6 Steel bearing piles

B6.1 General

All materials and work shall be in accordance with Sections B1 and B6 of this Specification, BS EN 12699 and BS EN 1993-5.

B6.2 Project Specification

The following matters are, where appropriate, described in the Project Specification:

(*a*) penetration or depth or toe level

(*b*) driving resistance or dynamic evaluation or set

(*c*) uplift/preliminary displacement trials

(*d*) pre-boring, jetting or other means of easing pile driving

(*e*) detailed requirements for driving records

(*f*) grades of steel

(*g*) sections of proprietary types of pile

(*h*) minimum length of pile to be supplied

(*i*) head and toe strengthening

(*j*) pile shoes (where required)

(*k*) practical refusal for pile extraction

(*l*) surface preparation

(*m*) types of coating

(*n*) thickness of primer and coats

(*o*) adhesion tests

(*p*) welding procedure — additional requirements to Clause B6.6

(*q*) non-destructive testing of welds — additional requirements to Clauses B6.6.3 and 6.6.5.3

(*r*) concreting of piles

(*s*) marking of piles — additional requirements to Clause 6.4.2

(*t*) other particular technical requirements.

B6.3 Materials

B6.3.1 Compliance with standards

The materials of steel piles shall as a minimum comply with BS 4, BS EN 10024, BS EN 10025, BS EN 10210, BS EN 10219 and BS EN 10034.

Fabrication of steel piles shall comply with BS EN 12699 and BS EN 1993-5.

B6.3.2 Inspection and test certificates

Where required to prove compliance with Clause 6.3, the Contractor shall provide the Engineer with test certificates for all steel piles to be incorporated into the Works prior to installation. The Contractor shall give the Engineer at least one week notice of the start of each stage of the manufacturing process and any production tests. The Contractor shall provide the Engineer with samples if specified.

B6.3.3 Pile shoes

Pile shoes shall comply with BS EN 12699.

B6.3.4 Manufacturing tolerances

All piles shall be of the type and cross-sectional dimensions as specified. The dimensional, length and weight tolerances shall comply with the relevant standard for the type of pile to be used.

B6.3.5 Straightness of sections

The deviation from straightness shall be within the compliance provisions of the relevant standard for the type of pile to be used.

When two or more sections are connected by butt-jointing by the manfacturer, the deviation from straightness shall not exceed 1/200 of the overall length of the pile.

B6.3.6 Fabrication of piles

For tubular piles where the load will be carried by the wall and where the pile will be subject to loads that induce reversal of stress during or after construction, the external diameter of any section as measured by using a steel tape on the circumference shall not differ from the theoretical diameter by more than ±1%.

The ends of tubular piles as manufactured shall be within a tolerance on ovality of ±1% as measured by a ring gauge for a distance of 100 mm at each end of the pile length.

The root edges or root faces of lengths of piles which are to be shop butt-welded shall not differ by more than 25% of the thickness of pile walls not exceeding 12 mm thick, or by more than 3 mm for piles where the wall is thicker than 12 mm. When piles of unequal wall thickness are to be butt-welded, the thickness of the thinner material shall be the criterion.

For site fabrication pile lengths shall be matched so that the differences in dimensions are distributed as evenly as possible.

B6.4 Construction processes
B6.4.1 Ordering of piles

Piles shall be ordered by the Contractor to meet programme requirements. All piles and production facilities shall be made available for inspection. Piles shall be examined by the Contractor at the time of delivery and any faulty piles rejected.

The Contractor shall notify the Engineer if time available from provision of all preliminary pile results to ordering of piles will be less than one week.

B6.4.2 Marking of piles

Each pile shall be labelled or marked to clearly indicate section size, steel grade and length.

In addition, if required by the Project Specification, each pile shall be graduated along its length at intervals of 250 mm before being driven.

B6.4.3 Handling and storage of piles

All piles shall where practicable be stacked in groups of the same section size, steel grade and length. Piles shall be placed on supports that are of sufficient size and frequency to prevent distortion. Stacks of piles shall be stable and spaced so that they are readily accessible. All operations such as handling, transporting and pitching of piles shall be carried out in a manner such that no significant damage occurs to piles and their coatings.

Minor damage to piles shall be repaired in accordance with the manufacturer's instructions. Damage to coatings shall be repaired in accordance with Clause B6.5.9.

B6.4.4 Driving piles

Pile installation and procedures shall be in accordance with BS EN 12699 and Clause B1.14.

B6.4.4.1 Steel screw piles

Screw piles shall be installed to the specified resistance or torque in accordance with the Contractor's method statement.

B6.4.5 Pile strengthening

If required by the Project Specification, strengthening to the toe of a pile in lieu of a shoe, or the strengthening of the head of a pile, shall be made using material of the same grade or higher than the pile.

B6.4.6 Pile installation system

The pile shall be safely supported by the pile installation system.

B6.4.7 Performance of driving equipment

The pile installation system shall be capable of adjustment to provide an axial, non-eccentric application of the installation force.

B6.4.8 Length of piles

The length of each pile driven shall be recorded, and any significant variation in length compared to the design length shall be reported to the Engineer within 24 hours.

B6.4.9 Matching of tubular pile lengths

If a hollow section pile is to be extended by butt welding at the joint, longitudinal or spiral seam welds used to form the sections

Specification for piling and embedded retaining walls. Thomas Telford, London, 2007

shall be evenly spaced around the pile perimeter. However, if a satisfactory match or the specified straightness can only be achieved by bringing the seams close together, they shall be offset by at least 100 mm.

B6.4.10 Matching of proprietary sections

When sections of proprietary piles are to be butt welded together, the sections to be joined shall be matched to achieve a satisfactory fit of webs and flanges. The acceptance criteria for a satisfactory match shall be defined in the Project Specification, taking into account the positon of the joint in the service condition and the effects of all loads acting on the pile at this position.

B6.4.11 Preparation of pile heads

If a steel structure is to be welded to piles, the piles shall be cut to within 10 mm of the levels specified.

If pile heads are to be encased in concrete they shall be driven to, or trimmed to, within 20 mm of the levels specified. Protective coatings shall be removed from the surfaces of the pile heads down to a level 100 mm above the soffit of the concrete.

B6.4.12 Extraction

Unless stated otherwise in the Project Specification, practical refusal for pile extraction shall be when the rate of extraction of a pile falls below 100 mm per minute when back-hammering or pulling (with equipment normally capable of withdrawing a pile) continuously (after an initial effort of 10 minutes), or when damage to the pile head occurs, or when the extraction process becomes unsafe.

B6.5 Coating piles for protection against corrosion

Where coatings are specified they shall be provided in accordance with the Project Specification. Piles shall not be delivered to site until the coatings have fully cured.

B6.5.1 Definition

The term 'coating' shall include the primer and all coats specified.

B6.5.2 Specialist labour

The preparation of surfaces and the application of the coats to form the coating shall be carried out safely by competent personnel.

B6.5.3 Protection during coating

All work associated with surface preparation and coating shall be undertaken inside a waterproof structure.

B6.5.4 Surface preparation

All surfaces to be coated shall be clean and dry and prepared by one or both of the following methods:

(a) Degreasing with detergent wash compatible with the coating shall be carried out where necessary.
(b) All surfaces shall be blast cleaned to BE EN ISO 8501-1 and BS 7079-A1. Blast-cleaning shall be done after fabrication. Unless an instantaneous-recovery blasting machine is used, the cleaned steel surface shall be air-blasted with clean dry air and vacuum-cleaned or otherwise freed from abrasive residues and dust immediately after cleaning.

B6.5.5 Application and type of primer

The surface shall be coated with an appropriate primer or the specified coating within 4 hours after surface preparation. No coating shall be applied to a metal surface which is not thoroughly dry.

The primer shall be compatible with the specified coating and shall be such that if subsequent welding or cutting is to be carried out it shall not emit noxious fumes or be detrimental to the welding.

B6.5.6 Control of humidity during coating

The method of application shall comply with the manufacturer's limits on humidity. Humidity shall be measured by a hygrometer or similar equipment.

B6.5.7 Parts to be welded

The coating within 200 mm of a planned weld shall be applied after welding. The system and method of application shall comply with the manufacturer's instructions.

Where attachments are to be welded to the piles after installation it will be necessary to ensure that the coating is removed and replaced where appropriate.

B6.5.8 Thickness, number and colour of coats

The nominal thickness of the finished coating and, if necessary, of each coat shall be as specified. The average coat or finished coating thickness shall be equal to or greater than the specified nominal thickness in accordance with BS EN ISO 12944. In no case shall any coat or finished coating be less than 75% of the nominal thickness. Each coat shall be applied after an interval that ensures the proper hardening or curing of the previous coat in accordance with the manufacturer's instructions.

Where more than one coat is applied to a surface, each coat shall, if possible, be of a different colour from the previous coat. The colour sequence and final coating shall be specified by the manufacturer.

B6.5.9 Inspection of coatings

The finished coating shall be generally smooth, of dense and uniform texture and free from sharp protuberances or pin-holes. Excessive sags, dimpling or curtaining shall be re-treated.

Any coat damaged by subsequent processes, or which have deteriorated to an extent such that proper adhesion of the coating is in doubt, shall be removed and the surface cleaned to the original standard and recoated to provide the original number of coats or an approved manufacturer's repair system used.

The completed coating shall be checked for thickness. Areas where the thickness is less than that specified shall receive additional treatment.

If specified in the Project Specification, the completed coating shall be checked for adhesion by means of an adhesion test to BS EN ISO 2409 and BS 3900-E6. The adhesion of any completed coating shall not be worse than Classification 2 of the Standards. Adhesion tests shall not be carried out until seven days after coating has cured. On completion of testing the test area shall be made good. Areas where the adhesion is defective shall be repaired and reinspected.

B6.6 Welding procedures

Testing and inspection of welds on site shall be in accordance with Table 1 of BS EN 12699, unless otherwise specified by the Project Specification. All procedures shall be in accordance with BS EN ISO 15607, BS EN ISO 15609-1, BS EN ISO 15613 and BS EN ISO 15614-1. The Contractor shall submit full details of the welding procedures and electrodes, with drawings and schedules, in the method statement.

Tests shall be undertaken in accordance with the relevant Standards and the Project Specification.

B6.6.1 Site welding

All welding shall be undertaken to BS EN 1011. Welders shall be qualified to BS EN 287-1 and proof of proficiency shall be submitted to the Engineer one week before commencement of the Works.

B6.6.2 Off-site welding
B6.6.2.1 Welded tubular piles

All fabrication drawings shall be submitted to the Engineer at least one week before the commencement of fabrication.

In addition, the Contractor shall submit details of fabrication and welding procedures to the Engineer one week before commencement of fabrication.

B6.6.2.2 Welded box piles and proprietary sections

The Contractor shall submit details of the fabrication and welding procedures for welded box piles, or proprietary sections made up from two or more hot-rolled sections, to the Engineer one week before commencement of fabrication.

Proprietary sections made up from two or more hot-rolled sections, shall be welded in accordance with the manufacturer's quality assurance programme.

B6.6.3 Non-destructive testing of welds

During production of welded tubular piles, one radiograph or ultrasonic test of a length of approximately 300 mm shall be made at each end of a length as manufactured to provide a spot check on weld quality. In addition, on a spirally welded tubular pile, a further check shall be made on welded joints between strip lengths.

All circumferential welds shall be fully radiographed or ultrasonically tested.

Results shall be submitted to the Engineer within one week of completion of the tests. If the results of any weld test do not conform to the specified requirements, two additional samples from the same length of pile shall be tested. In the case of failure of one or both of these additional tests, the length of pile covered by the tests shall be rejected.

B6.6.4 Standards for welds
B6.6.4.1 Longitudinal welds in tubular piles

For piles of longitudinal or spiral weld manufacture where the load will be carried by the wall of the pile, and if the pile will be subject to loads which induce reversal of stress during or after construction other than driving stresses, the standard for interpretation of non-destructive testing shall be the American Petroleum Institute Specification 5L. The maximum permissible height of weld reinforcement shall not exceed 3.2 mm for wall thicknesses not exceeding 12.7 mm and 4.8 mm for wall thicknesses greater than 12.7 mm.

B6.6.4.2 Longitudinal welds in proprietary box piles and proprietary sections

Longitudinal welds joining the constituent parts of the box or proprietary section shall be in accordance with the manufacturer's specification.

B6.6.4.3 Circumferential welds

For circumferential welds in tubular piles, the same maximum height of weld reinforcement as specified in Clause B6.6.4.1 shall apply. The standard for interpretation of non-destructive testing shall be the American Petroleum Institute Specification 5L.

B6.6.5 Site-welded butt splices
B6.6.5.1 Support and alignment of sections

Adequate facilities shall be provided for supporting and aligning the pile sections to be joined.

In the case of piles which are extended after driving an initial section into the ground, the alignment of the two sections shall be measured over a distance of 500 mm either side of the joint (1000 mm in total). The critical plane of alignment (flange or web) shall be identified in the Project Specification and the maximum permissible deviation from axial alignment shall be 2% of the width of the section plus an allowance for the rolling tolerances on straightness.

B6.6.5.2 Standards for site welds

Unless otherwise specified by the Project Specification, all welds shall comply with the requirements of BS EN 12699.

B6.6.5.3 Weld tests

Weld tests in accordance with BS EN 5817 shall be performed by ultrasonic or radiographic methods and as required by the Project Specification. Provided that satisfactory results are obtained, one test of a length of 300 mm shall be made for 10% or more of the number of welded splices in the case where the load will be carried by the wall or section of the pile. Where the load will be carried by the concrete core of the pile, the number of tests shall be specified by the Engineer, but shall not exceed 10% of the number of butt splices.

Results shall be submitted to the Engineer within one week of completion of the tests. Any defective welds shall be removed, replaced and reinspected.

B6.6.5.4 Protection during welding

All work associated with welding shall be protected from the weather so that the quality of the work meets the requirements of the Specification.

B6.7 Records

All records shall be in accordance with Clause B1.12 and the Project Specification.

B7 Timber piles
B7.1 General

All materials and work shall be in accordance with BS EN 1995-1-1, BS EN 12699 and Sections B1 and B7 of this Specification.

B7.2 Project Specification

The following matters are, where appropriate, described in the Project Specification:

(*a*) penetration or depth or toe level
(*b*) driving resistance or dynamic evaluation or set
(*c*) uplift/lateral displacement trials
(*d*) pre-boring, jetting or other means of easing piling driveability
(*e*) detailed requirements for driving records
(*f*) grades and types of piles shoes
(*g*) species and grades of timber
(*h*) preservative treatment
(*i*) certification requirements
(*j*) details of pile encasement where required
(*k*) splicing
(*l*) lengths and dimensions
(*m*) other particular technical requirements.

B7.3 Materials
B7.3.1 Species/grade of timber

The species of timber shall be as stated in the Project Specification. Timber grading shall comply with the general requirements stated within BS EN 14081-1.

The grade of timber shall be as follows:

- softwoods — 'special structural' (SS) grade as defined in BS 4978
- tropical hardwoods — grade HS as defined in BS 5756
- temperate hardwoods — grade THA as defined in BS 5756

or equivalent grades as defined by the Project Specification.

B7.3.2 Sapwood

Tree trunks for use as round piles shall have the bark removed but the sapwood left in place. They shall be treated with preservative as specified. Sawn or hewn softwood or hardwood that is to be treated with preservative need not have the sapwood removed. Hardwood that is to be used untreated shall be free of sapwood.

B7.3.3 Tolerance in timber dimensions

The tolerance on cross-sectional dimensions of sawn piles in relation to the dimensions specified shall be between +12 mm and −6 mm. The centroid of any cross-section of a sawn pile shall not deviate by more than 25 mm from the straight line connecting the centroids of the end faces of the pile.

Hewn piles shall be evenly tapered. The section dimensions shall not change more than 15 mm/m. The straightness of the pile shall not deviate from the straight line by more than 1% of the length.

B7.3.4 Preservatives

All preservation products shall comply with the Project Specification and be applied in accordance with the manufacturer's instructions.

B7.3.5 Certification of timber

Confirmation shall be obtained that the timber has been sourced from well-managed forests or plantations in accordance with the laws governing forest management in the producer country, together with any other certification required in the Project Specification.

B7.3.6 Pile shoes

Pile shoes shall comply with BS EN 12699 and be manufactured from durable material capable of withstanding the stresses caused by the installation methods and ground conditions without damage. The material and dimensions of the pile shoes shall be as specified.

Cast-iron shoes shall be made from chill-hardened iron as used for making grey iron castings to BS EN 1561, Grade EN-GJL-250, or ductile iron in accordance with BS EN 1563. The chilled iron point shall be free from major blow-holes and other surface defects.

Cast-steel pile shoes shall be fabricated from steel to BS EN 10293, Grade GS-240.

Fabricated shoes and their fastenings shall be made from steel to BS EN 10025, Grade S275 or S355.

Straps and other fastenings to cast pile shoes shall be of steel to BS EN 10025, Grade S275, and shall be cast into the point to form an integral part of the shoe.

B7.4 Construction processes
B7.4.1 Inspection and stacking

The Contractor shall notify the Engineer of the delivery of timber piles to the site or to the place of preservative treatment, and provide all labour and materials to enable the Engineer to inspect each piece on all faces and to measure it at the time of unloading and immediately prior to driving.

Timber shall be marked and stacked in lengths on paving or drained hard ground. Each piece of timber shall be clear of the ground and have an air space around it. The baulks or piles shall be separated by suitable blocks or spacers placed vertically one above the other and positioned at centres which are close enough to prevent sagging. The timber shall be protected from the weather by means of roofing over with tarpaulins or other appropriate covering which allows free circulation of air.

B7.4.2 Treatment with preservative

Preservative treatment shall be carried out in accordance with the recommendations of BS 8417, BS 144 or BS 4072 as specified. Cutting and boring of timber shall be done as far as possible before preservative treatment, but, where this is not possible, all surfaces subsequently cut or bored shall be heavily coated with preservative as specified in the relevant British Standard for preservative treatment or in accordance with the manufacturer's instructions as appropriate.

Certificates of treatment must be obtained and presented to the Engineer for all treated timber. The type and method of treatment must be compatible with the type of timber and the use to which the timber so treated is to be put.

B7.4.3 Pile shoes

The shoes shall be attached to the pile by steel straps fixed, spiked, screwed or bolted to the timber. The shoes shall be coaxial with the pile and firmly bedded to it.

B7.4.4 Pile heads

The pile head shall be flat and at right angles to the axis of the pile.

Before driving, precautions to prevent brooming shall be taken. This may be done by trimming the head of the pile square to the axis and fitting it with a steel or iron ring. The ring shall be not less than 50 mm deep by 12 mm thick in cross-section and the join shall be welded for its full section. The external diameter of the ring shall be that of the least allowable transverse dimension of the head of the pile. The top of the ring shall be between 10 mm and 20 mm from the top of the pile. If the ring is displaced during driving it shall be refitted. If the ring is broken a new ring shall be fitted.

As an alternative to a ring, a metal helmet may be used, the top of the pile being trimmed to fit closely into the recess of the underside of the helmet. A hardwood dolly and, if necessary, a packing shall be used above the helmet.

If during driving the head of the pile becomes excessively broomed or otherwise damaged, the damaged part shall be cut off, the head retrimmed and the ring or helmet refitted.

After driving, the heads of the piles should be cut off square to sound wood and treated with preservative before capping.

B7.4.5 Splicing

Piles shall be provided in one piece unless otherwise specified. A splice shall be capable of resisting safely any stresses which may develop during lifting, pitching or driving, and under loading. The position and details of the splice shall be as specified.

The splice shall be made as follows. The two timbers shall be of the same sectional dimensions and each cut at right angles to its axis to make contact over the whole of the cross-sectional when the two timbers are coaxial. A jointing compound shall be used at the contact surface. Round timbers shall be joined by a section of steel tube. Rectangular piles shall be joined by a prefabricated steel box section fitting the timbers closely or by steel splice plates. The connection shall be bolted, screwed or spiked to the timbers to keep the joined ends in close contact. The two parts shall not be more than 1:100 out of axial alignment.

Where it is necessary to extend a partly driven pile, the upper part must be securely supported during the making of the joint.

B7.4.6 Length of piles

The length of pile supplied to be driven in any position and any additional lengths to be added during driving shall comply with the Specification. During the execution of the Works any changes to the supplied lengths shall be made known to the Engineer within 24 hours.

B7.4.7 Spliced piles

Spliced piles shall be observed continuously during driving to detect any departure from true alignment of the two parts. If any such departure occurs, driving shall be suspended and the Engineer shall be informed within 24 hours.

B7.4.8 Preparation of pile heads

After driving, the piles shall be cut off square to sound timber to within 5 mm of the levels shown on the drawings and the cut surfaces shall be heavily coated with preservative as specified for the original treatment.

B7.4.9 Driving piles

Pile installation and procedures shall be in accordance with BS EN 12699 and Clause B1.14.

B7.5 Records

All records shall be in accordance with the requirements of Clause B1.12 and the Project Specification.

B8 Diaphragm walls and barrettes

B8.1 General

All materials and work shall be in accordance with Sections B1, B8, B18 and B19 of this Specification and BS EN 1538.

In this section the term panel shall refer to both walls and barrettes.

B8.2 Project Specification

The following matters are, where appropriate, described in the Project Specification:

(a) construction tolerances (if tolerances more onerous than those in Table B1.4 are required)
(b) performance criteria for movement under vertical loads
(c) support fluid
(d) panel dimensions (minimum or maximum thickness and/or panel length)
(e) additional overbreak tolerance
(f) preliminary barrettes
(g) water stop requirements (diaphragm walls only)
(h) instrumentation
(i) base or shaft grouting
(j) temporary backfill material
(k) integrity testing
(l) permissible materials for permanent stop-ends
(m) period for Engineer approval of Contractor's drawings
(n) other particular technical requirements.

B8.3 Materials

B8.3.1 Support fluid

Bentonite and alternative support fluid materials, additives, mixing, testing, and use of clean water, shall be in accordance with Section B18.

B8.3.2 Permanent stop-ends in diaphragm wall panels

Materials shall be in accordance with the Project Specification.

B8.3.3 Grout

All requirements with regard to shaft or base grouting shall be in accordance with Section B19.

B8.3.4 Alternatives to steel reinforcement

The Contractor shall submit with the tender information on the material properties.

B8.4 Construction tolerances

B8.4.1 Guide wall

The finished internal face of the guide wall closest to any subsequent main excavation shall be vertical to within a tolerance of 1 in 200 and the top edge of the wall shall represent the reference line. There shall be no ridges or abrupt changes on the face and its variation from its specified position shall not exceed ±15 mm in 3 m.

The minimum clear distance between the guide walls shall be the specified diaphragm wall thickness plus 25 mm and the maximum distance shall be the specified diaphragm wall thickness plus 50 mm.

B8.4.2 Diaphragm wall and barrettes

Tolerances shall be in accordance with Clause B1.8, unless otherwise stated in the Project Specification. In the case of barrettes, the above specified maximum deviation shall apply in any direction.

An additional tolerance of 100 mm shall be allowed for concrete protrusions formed by overbreak or resulting from cavities in the ground. Where soft, loose or organic soil layers are anticipated, or obstructions are to be removed during panel excavation, an additional overbreak tolerance shall be stated in the Project Specification. Responsibility for the removal of protrusions shall be stated in the contract.

B8.4.3 Recesses

Where recesses are to be formed by inserts in the wall, the vertical tolerance shall be that of Clause B8.4.4, and the horizontal tolerance shall be that of Clause B8.4.4 plus the horizontal tolerance from Clause B8.4.2.

B8.4.4 Reinforcement

Reinforcement shall be maintained in its correct position during concreting of the panel within a vertical tolerance of $+150/-50$ mm (i.e. a maximum of 150 mm high) on the level of the reinforcement projecting above the final cut-off level.

The longitudinal tolerance of the cage at final cut-off level measured along the excavation shall be ±75 mm within the panel.

B8.4.5 Concrete casting level

If the cut-off level for the panel is less than 1 m below the top level of the guide walls, uncontaminated concrete shall be brought to the top of the guide walls. If the cut-off level is greater than 1 m below the top level of the guide walls, concrete shall be brought to 1 m above the cut-off level specified, with a tolerance of ±150 mm. An additional tolerance of $+150$ mm over the above tolerances shall be permitted for each 1.0 m of depth by which the cut-off level is below the top of the guide wall, up to a maximum of 2.0 m.

B8.4.6 Dimensions of panels

The thickness of a panel shall be not less than the specified thickness.

The lengths of all panels shall be within the limits on length specified in the Project Specification. The Contractor shall be responsible for selecting panel dimensions which ensure stability. If, in the Contractor's opinion, the specified panel dimensions are not adequate to ensure stability, he shall inform the Engineer at the time of tender.

B8.4.7 Water retention

The Contractor shall carry out the repair of any joint, defect or panel where, on exposure of the diaphragm wall, the wall is not deemed 'watertight' as specified in Clause B1.9.

B8.5 Construction processes
B8.5.1 Drawings

The Contractor shall produce construction drawings including a panel layout drawing and reinforcement fabrication drawings for each panel. These drawings shall be submitted to the Engineer in accordance with the approval period stated in the Project Specification.

B8.5.2 Guide walls

The design and construction of the guide walls shall be the responsibility of the Contractor and shall take into account the actual site and ground conditions, all temporary loadings whether from reinforcement cages, stop-ends or equipment and the required level of support fluid to ensure stability and avoid undercutting of the guide walls. Guide walls shall be constructed in reinforced concrete or other suitable materials. The minimum depth of guide walls shall be 1.0 m and the minimum shoulder width shall be 0.3 m for reinforced concrete guide walls. Reinforcement continuity shall be provided between adjacent sections of guide walls.

B8.5.3 Stability of the excavation

A suitable guide wall shall be used in conjunction with the method to ensure stability of the strata near ground level until concrete and any backfill material has been placed.

During construction the level of support fluid in the excavation shall be maintained within the guide wall or stable ground so that stability is ensured.

In the event of a loss of support fluid from an excavation, the Contractor shall notify the Engineer of his intended action before proceeding.

B8.5.4 Cleanliness of the base

Prior to placing reinforcement or concrete, the Contractor shall clean the base of the excavation of loose, disturbed and remoulded materials in accordance with the method statement.

B8.5.5 Condition of support fluid prior to concreting

The Contractor shall ensure that contaminated support fluid, which could impair the free flow of concrete from the tremie pipe, has not accumulated in the bottom of the panel. The Contractor shall wholly or partly remove and replace contaminated support fluid whilst maintaining stability, until it complies with the stated limits for the support fluid.

B8.5.6 Stop-ends in diaphragm wall panels

The Contractor shall state in his tender the type of stop-end proposed and whether they will be removed or permanently cast into the panel.

Stop-ends shall be of the length, thickness and quality of material adequate for the purpose of preventing water and soil from entering the panel excavations. The external surface shall be clean and smooth.

Stop-ends shall be rigid and adequately restrained to prevent horizontal movement during concreting.

If excavation is by the use of a reverse circulation mill and stop-ends are not proposed, the Contractor shall provide sufficient overcutting of the adjacent panel to ensure that a continuous competent joint is produced over the full depth of the panel.

B8.5.7 Placing concrete

Before commencement of concreting of a panel, the Contractor shall satisfy himself that the supplier will have available sufficient quantity of concrete to construct the panel in one continuous operation.

The concrete shall be placed without such interruption as would allow the previously placed batch to have achieved a stiffness which prevents proper amalgamation of the two concrete batches.

No spoil, liquid or other foreign matter shall be allowed to contaminate the concrete.

The slump range or target flow for concrete placed through support fluid using a tremie pipe shall be stated in the method statement.

Vibrators shall not be used to compact concrete within a cast-in-place panel.

The concrete shall be placed through a tremie pipe in one continuous operation. Where two or more tremie pipes are used in the same panel simultaneously, care shall be taken to ensure that the difference in concrete level at each pipe position is not greater than ±250 mm. In the case of walls or barrettes which are designed to carry a proportion of load in end bearing, on commencement of the pour concrete shall be discharged at each tremie simultaneously until complete coverage of the base of the panel has been achieved.

The hopper and pipe of the tremie shall be clean and watertight throughout. The tremie pipe shall extend to the base of the panel and a sliding plug or barrier shall be placed in the pipe to prevent direct contact between the first charge of concrete in the tremie and the support fluid.

The tremie pipe shall at all times penetrate the concrete which has previously been placed with a maximum embedment of 6 m and a minimum embedment of 3 m and shall not be withdrawn from the concrete until completion of concreting. At all times a sufficient quantity of concrete shall be maintained within the pipe to ensure that the pressure from it exceeds that from the support fluid and workable concrete above the tremie base.

The internal diameter of the pipe of the tremie shall be of sufficient size to ensure the easy flow of concrete. The internal face of the pipe of the tremie shall be free from projections.

The tremie pipe shall be so designed that external projections are minimized, allowing the tremie to pass within reinforcing cages without causing damage.

B8.5.8 Temporary backfilling above panel casting level

After each panel has been cast, any empty excavation remaining shall be protected and shall be carefully backfilled as soon as possible with inert material in accordance with the Project Specification. Prior to backfilling, panels shall be clearly marked and fenced off so as not to cause a safety hazard.

B8.5.9 Excavation near recently cast panels

Panels shall not be excavated so close to other panels which have recently been cast and which contain workable or unset concrete that a flow of concrete or instability could be induced or damage caused to any panel.

B8.5.10 Breaking down of concrete

When cutting off and breaking down concrete to specified cut-off levels, the Contractor shall take care to avoid shattering or otherwise damaging the concrete, reservation pipes and/or waterbars. Any cracked or defective concrete shall be cut away and repaired in a manner to provide a full and sound section at cut-off level.

B8.5.11 Base or shaft grouting

All procedures shall be in accordance with Clause B3.5.11.

B8.6 Records

All records shall be in accordance with the requirements of Clause B1.12 and the Project Specification.

B8.6.1 Placing concrete

The depth of the surface of the concrete shall be measured and the embedded length of the tremie pipe recorded at regular intervals corresponding to each batch of concrete. The depths measured and volumes placed shall be plotted immediately on a graph during the concreting process and compared with the theoretical relationship of depth against volume, and shall be provided to the Engineer within 24 hours.

B8.6.2 Records of special control measures during excavation

If installation tolerances are stated in the Project Specification that are more onerous than those in Clause B1.8, then the additional records needed to prove that these tolerances have been achieved shall be provided to the Engineer within 48 hours.

B9 Secant pile walls

B9.1 General

All materials and work shall be in accordance with Sections B1, B3, B4, B9, B18 and B19 of this Specification, and BS EN 1536.

B9.2 Project Specification

The following matters are, where appropriate, described in the Project Specification:

(a) construction tolerances (if tolerances more onerous than those in Table B1.4 are required)
(b) performance criteria for movement under vertical loads
(c) support fluid
(d) additional overbreak tolerance
(e) requirements for self-hardening slurry mixes such as strength, permeability, shrinkage and durability
(f) pile diameters
(g) pile spacing and overlap at commencing level
(h) depth to which pile interlock must be maintained
(i) instrumentation
(j) temporary backfill material
(k) integrity testing
(l) other particular technical requirements.

B9.3 Materials

B9.3.1 Self-hardening slurry mixes and low-strength concrete mixes

The Contractor shall submit details for the self-hardening slurry mix or for the low-strength concrete mix proportions to be used prior to commencing the Works.

Cement materials, aggregates (if used), additives and water shall be in accordance with Section B19.

Trial mixes shall be prepared for each mix unless there is existing data showing that the mix proportions and method of manufacture will produce hardened material of the strength, permeability, shrinkage properties and durability required, having adequate workability for compaction by the method to be used in placing. The performance requirements are to be stated in the Project Specification. The requirements for testing shall be set out in the Project Specification and samples shall be tested in a third party accredited laboratory for these tests.

Where a trial mix is required after commencement of the Works the above procedure shall be adopted. The workability of each batch of a trial mix shall be determined by the method as specified in the Project Specification.

No variations in the original source or properties of the materials outside the specified limits shall be made without demonstrating compliance with this Specification.

Mixes shall be checked for compliance with the mix proportions. Cylinder or cube samples shall be made at the rate of one set of four specimens for each $50\,\text{m}^3$ or part thereof in each day's work. Testing shall be in accordance with the requirements of the Project Specification and carried out in a third party accredited laboratory for these tests. The results shall comply with the Project Specification.

The Contractor shall keep a detailed record of the results of all tests on self-hardening mixes or for the low-strength concrete mixes and their ingredients. Each test shall be clearly identified with the pile to which it relates and the date it was carried out.

B9.3.2 Support fluid

Where support fluid is used to maintain the stability of the pile bore it shall be in accordance with Section B18.

B9.3.3 Alternatives to steel reinforcement

The Contractor shall submit with the tender information on the material properties.

B9.4 Construction tolerances

B9.4.1 Guide wall

The finished internal face of the guide wall closest to any subsequent main excavation shall be vertical to a tolerance of 1 in 200 and shall represent the reference line. There shall be no ridges on this face and the centre line of the guide wall shall not deviate from its specified position by more than +15 mm/−15 mm in any 3 m along its length.

The clear distance between the inside faces of the guide wall shall be the maximum tool diameter plus 25 mm, with a tolerance of +25/−0 mm.

B9.4.2 Secant piles

Tolerances shall be in accordance with Clause B1.8, unless otherwise stated in the Project Specification. An additional tolerance of 100 mm shall be allowed for concrete protrusions formed by overbreak or resulting from cavities in the ground. Where soft, loose or organic soil layers are anticipated, or obstructions are to be removed during pile excavation, an additional overbreak tolerance shall be stated in the Project Specification. Responsibility for the removal of protrusions shall be stated in the contract.

B9.4.3 Recesses

If recesses in the form of box-outs are to be formed within a pile shaft, the vertical tolerance shall be in accordance with Clause B9.4.4 and rotational tolerance shall be 10 degrees.

B9.4.4 Reinforcement

On completion of the pile, the reinforcement shall be located within a vertical tolerance of +150/−50 mm (i.e. a maximum of 150 mm high) on the level of the reinforcement projecting above the final cut-off level.

B9.4.5 Water retention

The Contractor shall carry out the repair of any joint, defect or pile where on exposure of the secant pile wall, the wall is not deemed 'watertight' as specified in Clause B1.9.

B9.5 Construction processes

B9.5.1 Guide walls

Scalloped guide walls shall be used for all secant pile walls that form part of the permanent works of the structure but may be omitted for temporary works. In all cases, the Contractor shall demonstrate by experience, calculation and monitoring on site that adjacent piles will remain interlocked to the depth specified in the Project Specification.

The design and construction of guide walls shall be the responsibility of the Contractor and shall take into account the actual site and ground conditions and the equipment to be used on site to ensure stability and avoid under-cutting as appropriate.

Guide walls shall be constructed in reinforced concrete or of other suitable materials. The minimum depth of guide wall shall be 0.5 m and the minimum shoulder width shall be 0.3 m for walls in reinforced concrete.

B9.5.2 Bored cast-in-place piles

Bored cast-in-place piles shall be constructed in accordance with Section B3.

B9.5.3 Continuous flight auger piles

Continuous flight auger piles shall be constructed in accordance with Section B4.

B9.5.4 Boring into recently cast piles

Piles shall not be bored into other piles which have recently been cast and which contain workable or partially set concrete that a flow of concrete or instability could be induced or damage caused to any installed piles.

B9.5.5 Placing concrete, low-strength concrete and self-hardening slurry mixes

Concrete, low-strength concrete and self-hardening slurry mixes shall be placed in accordance with Sections B3 or B4, as appropriate.

B9.5.6 Breaking down of concrete

When cutting off and breaking down concrete to specified cut-off levels, the Contractor shall take care to avoid shattering or otherwise damaging the concrete and/or reservation pipes. Any cracked or defective concrete shall be cut away and repaired in a manner to provide a full and sound section at cut-off level.

B9.6 Records

All records shall be in accordance with the requirements of Clause B1.12 and the Project Specification.

B10 Contiguous pile walls

B10.1 General

All materials and work shall be in accordance with Sections B1, B3, B4, B10, B18 and B19 of this Specification, and BS EN 1536.

B10.2 Project Specification

The following matters are, where appropriate, described in the Project Specification:

(a) construction tolerances (if tolerances more onerous than those in Table B1.4 are required)
(b) performance criteria for movement under vertical loads
(c) support fluid
(d) additional overbreak tolerance
(e) pile diameters
(f) pile spacing at commencing level
(g) instrumentation
(h) temporary backfill material
(i) integrity testing
(j) other particular technical requirements.

B10.3 Materials
B10.3.1 Support fluid

Where support fluid is used to maintain the stability of the pile bore, it shall be in accordance with Section B18.

B10.3.2 Alternatives to steel reinforcement

The Contractor shall submit with the tender information on the material properties.

B10.4 Construction tolerances
B10.4.1 Guide wall

In cases where the use of a guide wall is specified, the finished internal face of the guide wall closest to any subsequent main excavation shall be vertical to a tolerance of 1 in 200 and shall represent the reference line. There shall be no ridges on this face and the centre line of the guide wall shall not deviate from its specified position by more than +15/−15 mm in any 3 m along its length.

The clear distance between the inside faces of the guide wall shall be the maximum tool diameter plus 25 mm, with a tolerance of +25/−0 mm.

B10.4.2 Contiguous piles

Tolerances shall be in accordance with Clause B1.8, unless otherwise stated in the Project Specification. An additional tolerance of 100 mm shall be allowed for concrete protrusions formed by overbreak or resulting from cavities in the ground. Where soft, loose or organic soil layers are anticipated, or obstructions are to be removed during pile excavation, an additional overbreak tolerance shall be stated in the Project Specification. Responsibility for the removal of protrusions shall be stated in the contract.

B10.4.3 Recesses

If recesses in the form of box-outs are to be formed within a pile shaft, the vertical tolerance shall be in accordance with Clause B10.4.4 and rotational tolerance shall be 10 degrees.

B10.4.4 Reinforcement

On completion of the pile the reinforcement shall be located within a vertical tolerance of +150/−50 mm (i.e. a maximum of 150 mm high) on the level of the reinforcement projecting above the final cut-off level.

B10.5 Construction processes
B10.5.1 Guide walls

Where the use of guide walls is specified, the design and construction of guide walls shall be the responsibility of the Contractor and shall take into account the actual site and ground conditions, and the equipment to be used on site to ensure stability.

Guide walls shall be constructed in reinforced concrete or of other suitable materials. The minimum depth of guide wall shall be 0.5 m and the minimum shoulder width shall be 0.3 m for walls in reinforced concrete.

B10.5.2 Bored cast-in-place piles

Bored cast-in-place piles shall be constructed in accordance with Section B3.

B10.5.3 Continuous flight auger piles

Continuous flight auger piles shall be constructed in accordance with Section B4.

B10.5.4 Placing concrete

Placing concrete shall be carried out in accordance with Sections B3 or B4, as appropriate.

B10.5.5 Breaking down of concrete

When cutting off and breaking down concrete to specified cut-off levels, the Contractor shall take care to avoid shattering or otherwise damaging the concrete and/or reservation pipes. Any cracked or defective concrete shall be cut away and repaired in a manner to provide a full and sound section at cut-off level.

B10.6 Records

All records shall be in accordance with the requirements of Clause B1.12 and the Project Specification.

Specification for piling and embedded retaining walls. Thomas Telford, London, 2007

B11 King post walls
B11.1 General

All materials and work shall be in accordance with Sections B1, B3, B4, B6, B11, B18 and B19 of this Specification, and BS EN 1536.

B11.2 Project Specification

The following matters are, where appropriate, described in the Project Specification:

(*a*) pile construction tolerances (if tolerances more onerous than those in Table B1.4 are required) and king post installation tolerances
(*b*) performance criteria for movement under vertical loads
(*c*) support fluid
(*d*) steel grade and sectional properties
(*e*) coating details
(*f*) pile diameter
(*g*) spacing
(*h*) instrumentation
(*i*) temporary backfill material
(*j*) whether it is permitted that king posts can be plunged into fluid concrete
(*k*) other particular technical requirements.

B11.3 Materials
B11.3.1 King post

The Contractor shall provide full details of the structural steel section to be employed at tender stage, including calculations showing how the structural steel section will achieve the requirements of the Project Specification. The structural steel section shall be in accordance with the relevant Standards.

B11.3.2 Support fluid

Where support fluid is used to maintain the stability of the pile bore it shall be in accordance with Section B18.

B11.4 Construction tolerances
B11.4.1 Piles

Tolerances shall be in accordance with Clause B1.8, unless otherwise stated in the Project Specification.

B11.4.2 King posts

On completion of the pile the section shall be located within a vertical tolerance of $+150/-150$ mm and with a rotational tolerance of 20 degrees unless otherwise stated in the Project Specification.

B11.5 Construction processes
B11.5.1 Bored cast-in-place piles

Construction of king post walls employing bored cast-in-place piling methods shall be constructed in accordance with Section B3.

B11.5.2 Continuous flight auger piles

Construction of king post walls employing continuous flight auger piling methods shall be constructed in accordance with Section B4.

B11.5.3 Placing king posts

The method of handling and placing king posts shall be set out in the Contractor's method statement.

Unless not permitted in the Project Specification, the king post may be plunged into concrete already placed in the pile bore immediately after the concrete placement and before the concrete has achieved its initial set. Guides or spacers shall be used so that the specified tolerances of the king post are achieved.

Where the king post is installed prior to concreting, the Contractor shall ensure that it is not disturbed, displaced or disoriented during concreting.

B11.5.4 Placing concrete

Concrete, low strength concrete and self-hardening slurry mixes shall be placed in accordance with Sections B3 or B4, as appropriate.

Where the king post is placed in the pile bore prior to concreting, a hopper with twin pipes shall be used and concrete will be placed in such a manner that a balanced concrete level is maintained on both sides of the section.

B11.5.5 Backfilling

The pile bore containing the king post shall be backfilled once the concrete has sufficient strength or after such time as specified in the Project Specification.

The backfill material shall be evenly placed on either side of the king post to avoid uneven loading. Prior to backfilling, bores shall be clearly marked and fenced off so as not to cause a safety hazard.

B11.6 Records

All records shall be in accordance with the requirements of Clause B1.12 and the Project Specification.

Specification for piling and embedded retaining walls. Thomas Telford, London, 2007

B12 Steel sheet piles
B12.1 General

All materials and work shall be in accordance with Sections B1 and B12 of this Specification and BS EN 12063.

B12.2 Project Specification

The following matters are, where appropriate, described in the Project Specification:

(a) construction tolerances (if tolerances more onerous than those in Table B1.4 are required)
(b) penetration or depth or toe level
(c) methods of installation or extraction
(d) detailed requirements for piling records
(e) grades of steel
(f) section type and manufacturer's reference number; minimum section modulus; second moment of inertia; web thickness
(g) surface preparation
(h) types of coating
(i) thickness of primer and coats
(j) cathodic protection
(k) types of head and toe preparation
(l) minimum length to be supplied
(m) practical refusal for pile installation or extraction
(n) interlock sealant or seal welds
(o) special piles, corners, etc.
(p) other particular technical requirements.

B12.3 Materials and fabrication
B12.3.1 Compliance with standards

Unless otherwise specified by the Project Specification, hot rolled sheet piles, king piles and connectors shall be manufactured to BS EN 10248, cold formed sheet piles to BS EN 10249, hot finished hollow sections to BS EN 10210 and cold finished hollow sections to BS EN 10219.

The dimensional tolerances of the piles shall comply with the relevant Standards above and BS EN 12063.

B12.3.2 Fabricated sheet piles

All fabricated piles, e.g. corners, junctions, box sections, high modulus sections, shall be fabricated and supplied in accordance with BS EN 12063. Where the Contractor proposes to use alternative fabricated piles to those detailed on the drawings approval of the Engineer shall be obtained before fabrication work commences.

B12.3.3 Inspection and test certificates

Where required to prove compliance with Clause B12.3 the Contractor shall provide the Engineer with test certificates for all steel piles to be incorporated into the Works prior to installation. The Contractor shall notify the Engineer at least one week before the start of the manufacturing process and any production tests. The Contractor shall provide the Engineer with samples if specified.

B12.3.4 Manufacturing tolerances

The dimensional, length and weight tolerances shall comply with the relevant Standard for the type of pile specified and to be used.

B12.3.5 Clutch sealant and seal welding

Sheet pile interlocks shall be sealed or welded in accordance with BS EN 12063, if required in the Project Specification. The Contractor shall supply piles with sealant applied in accordance with the manufacturer's instructions.

Details of the brand and properties of interlock sealant shall be supplied by the Contractor at tender. For temporary works the specification for sealing sheet pile interlocks shall be included in the Contractor's method statement.

B12.4 Construction processes

B12.4.1 Ordering of piles

Piles shall be ordered by the Contractor to meet programme requirements. All piles and production facilities shall be made available for inspection where practical. Piles, sealants and coatings shall be examined by the Contractor at the time of delivery to ensure compliance with B12.3.1 and any damaged piles shall be repaired or replaced.

Sheet piles shall be supplied in one length, unless otherwise permitted in the Project Specification, e.g. for low headroom applications. Spliced joints shall be designed to cater for the combined effects of bearing, shear and bending stresses imposed upon the piles. Splices shall be located to avoid maximum stress positions. If splices are to be welded, then these shall be designed to comply with BS EN 12063 and the manufacturer's instructions, and be staggered vertically by at least 500 mm between adjacent piles. Weld metal shall not encroach within the interlock areas so as not to interfere with the interlocking of the piles.

B12.4.2 Marking of piles

Each pile shall be labelled or marked to clearly indicate section size, steel grade and length.

B12.4.3 Handling and storage of piles

All piles shall where practicable be stacked in groups of the same section size, steel grade and length. Piles shall be placed on supports that are of sufficient size and frequency to prevent distortion. Stacks of piles shall be stable, on level ground, and spaced so that they are readily accessible. All operations such as handling, transporting and pitching of piles shall be carried out in a manner such that no significant damage occurs to piles and their coatings.

Minor damage to piles shall be repaired in accordance with the manufacturer's instructions. Damage to coatings shall be repaired in accordance with Clause B6.5.9.

When assembling piles before pitching, the Contractor shall ensure that the interlocks are clean and free from distortion.

B12.4.4 Pile installation

Sheet piles shall be installed safely and in accordance with BS EN 12063.

Sheet piles shall be installed within the tolerances given in Table 2 of BS EN 12063. For piles in excess of 20 m length, the tolerance on verticality shall be measured on the exposed height rather than the upper 1 m as defined in Table 2 of BS EN 12063.

The methods for access, handling, pitching and installing the piles, together with details of proposals to guide the piles to ensure verticality, shall be provided by the Contractor at tender stage. If pitch and drive methods are to be used, the length of the pile driven before pitching the next pile shall be stated. For panel driving and backdriving the maximum length of the lead on each pile or pair driven ahead of its neighbour shall be stated. The methods of installation and details of the equipment to achieve the final penetration levels shall be described. Pre-augering, jetting and other means of altering ground conditions to achieve final penetration levels shall not be permitted without the approval of the Engineer.

If pre-augering or jetting is proposed by the Contractor to aid installation, the depth applied and maximum pressures shall be stated in the Contractor's method statement.

Pile line dimensions shall be based on the nominal width and depth of the piles. Creep or shrinkage of the pile lines shall not exceed the manufacturer's rolling tolerances.

The Contractor shall be satisfied that the sheet piles can be installed adequately to the correct depths through the anticipated ground conditions.

Care shall be taken when pre-augering to ensure that the ground beneath the centreline of the sheet piles is not adversely affected and that the design of the sheet pile wall is not compromised at any stage of construction. Jetting shall only be carried out with the minimum pressure required to achieve installation of the sheet piles. Jetting shall not be permitted where there is a risk of causing damage.

Each pile shall be properly guided and fully interlocked with its neighbour. Piles shall not bypass or overlap one another in place of interlocking. Piles shall be prevented from rotation in excess of the maximum manufacturer's recommendations for the specified section.

Where sheet piles are installed in panels, the end piles to each panel shall be driven in advance of the general run of piles. After allowing for initial penetration, no pile in the panel shall be driven to an excessive lead in comparison with the toe level of the panel in general, and where hard driving is encountered this lead shall not exceed 1 m.

At all stages during installation the sheet pile shall be adequately supported and restrained in order to maintain verticality. The Contractor shall ensure that the sheet panels are driven without declutching or damage. The Contractor shall inform the Engineer of any declutching, damage or separation of the sheet pile wall, and submit proposals to the Engineer in accordance with Clause B1.16.

The selection of plant shall be commensurate with the ground conditions and pile type. Piles shall be driven to the specified level and/or resistance. If hard driving is experienced, piles shall be driven in pairs to practical refusal, defined as 12 blows per 25 mm penetration (with appropriate equipment) or when the rate of penetration is below 100 mm per minute when hammering continuously for no more than 5 minutes with double or single acting mechanical impact hammers. When driving piles into rock, penetration rates of up to 25 blows per 25 mm shall be permitted, but only for short durations in accordance with the Contractor's method statement. If practical refusal or visible damage to the pile head occurs during installation, work shall cease and the Engineer notified.

If the piles have not penetrated to the depth or level stated in the Project Specification, or have encountered obstructions, the Contractor shall submit proposals to the Engineer in accordance with Clause B1.16.

B12.4.5 Pile extraction

Methods and proposals for pile extraction shall be submitted to the Engineer for approval where required by the Project Specification.

Unless stated otherwise in the Project Specification, practical refusal for pile extraction shall be when the rate of extraction of a pile falls below 100 mm per minute when extracting or pulling continuously (after an initial effort of 10 minutes), or when damage to the pile head occurs, or when the extraction process becomes unsafe.

B12.4.6 Pile installation system

Piles shall be safely supported and aligned by the pile installation system.

Pile driving hammers shall be positioned on the pile so that the hammer will be aligned as near to the axis of the pile as is practically possible. Freely suspended piling hammers shall be equipped with correctly adjusted leg guides and inserts. Where a hammer is mounted in a rigid leader, the leader shall be stable. For impact driving, the anvil block or driving cap shall be of sufficient size to cover as much as possible of the full cross-section of the pile.

B12.4.7 Performance of installation equipment	Piling hammers and equipment shall be capable of adjustment to perform at maximum efficiency level for the type selected and provide an axial, non-eccentric application of the installation force.
B12.4.8 Length of piles	The length of each pile installed shall be recorded, and any significant variation to the specified length shall be reported to the Engineer within 24 hours.
B12.4.9 Methods of assisting pile installation	Any restrictions on the methods of assisting pile installation, such as the use of water jetting, shall be stated in the Project Specification. Methods to be used, including the use, control, treatment and disposal of water for jetting, shall be included in the Contractor's method statement.
B12.4.10 Preparation of pile heads	If a steel structure is to be welded to piles, the piles shall be cut to within 10 mm of the levels specified. If pile heads are to be encased in concrete, they shall be installed to, or trimmed to, within 20 mm of the levels specified. Protective coatings shall be omitted or removed from the surface of the pile heads down to a level 100 mm above the soffit of the concrete. For temporary works the pile heads shall be cut or installed to within 50 mm of the levels specified.
B12.5 Coating piles for protection against corrosion	All materials and work shall be carried out in accordance with BS EN ISO 12944 and Clause B6.5. Any treatment to the inside of interlocks shall be stated in the Project Specification.
B12.6 Welding procedures	Testing and inspection of welds on site shall be in accordance with Table 1 of BS EN 12063, unless otherwise stated in the Project Specification. All procedures shall be in accordance with BS EN ISO 15607, BS EN ISO 15609-1, BS EN ISO 15613 and BS EN ISO 15614-1. The Contractor shall submit full details of the welding procedures and electrodes, with drawings and schedules, in the method statement. Tests shall be undertaken in accordance with the relevant Standards and the Project Specification.
B12.6.1 Welding standards	Welders shall be qualified to BS EN 287-1 and proof of proficiency shall be submitted to the Engineer at least one week before commencement of the Works. For manual metal arc and semi-automatic welding of carbon and carbon manganese steels, welding of piles and steel framework shall be carried out in accordance with BS EN 1011-1 and BS EN 1011-2. Defective welds shall be removed and replaced. Where steel piles are to be spliced by butt welding, the interlocks shall not be welded unless a sealing weld is required. All butt welds shall comply with the requirements of BS EN 12063.
B12.6.2 Seal welds	Seal welds shall have a minimum thickness as stated in the Project Specification. Weld procedures shall be submitted to the Engineer for approval at least one week before commencement of welding. Category D acceptance level quality for the seal welds shall be required unless otherwise stated in the Project Specification. Visual examination and testing for at least 10% of the full length of seal welds shall be undertaken to prove that the welds are free from surface cracking and are watertight. Any repairs shall be carried out to a procedure approved by the Engineer and re-testing undertaken as appropriate.

Testing shall be carried out at least three days after the weld to be tested has been completed.

B12.6.3 Support and alignment

When lengths of pile are to be butt-spliced on site, adequate facilities shall be provided for supporting and aligning them prior to welding such that the deviation from straightness shall not exceed 1/200 of the overall length of the pile.

The Contractor shall supply details of corner fabrication for approval by the Engineer where creep and shrinkage has occurred during installation.

B12.6.4 Weld tests

Weld tests in accordance with BS EN 5817 shall be performed by ultrasonic or radiographic methods and as required by the Project Specification. Provided that satisfactory results are obtained, one test of a length of 300 mm shall be made for 10% or more of the number of welded splices.

Results shall be submitted to the Engineer within one week of completion of the tests. Any defective welds shall be removed, replaced and re-inspected.

B12.6.5 Protection during welding

All work associated with welding shall be protected from the weather so that the quality of the work meets the requirements of the Specification.

Site personnel and the general public shall be protected from welding operations.

B12.7 Records

All records shall be in accordance with Clause B1.12 and the Project Specification.

B13 Integrity testing

B13.1 General

This section is relevant to the testing of cast-in-situ piles, barrettes and wall elements for the integrity of the concrete.

B13.2 Method of testing

Where integrity testing is called for, the method to be adopted shall be one of the following, as specified:

    (a) impulse response method (piles and barrettes only)

    (b) sonic echo, frequency response or transient dynamic steady-state vibration method (piles and barrettes only)

    (c) cross-hole sonic logging method.

Other methods may be considered by the Engineer subject to satisfactory evidence of performance.

B13.3 Project Specification

The following matters are, where appropriate, described in the Project Specification:

    (a) the method of test to be carried out

    (b) the number, type and location of elements to be tested

    (c) the stages in the programme of works when a phase of integrity testing is to be carried out

    (d) where cross-hole sonic logging is used, the following shall be specified: the number and location of elements to be tested, the number and location and length of ducts to be placed, the depth over which the testing is required and the depth interval that shall not be greater than 0.25 m

    (e) the time after testing at which the test results and findings shall be available to the Engineer, if different from the requirements of Clause B13.8

    (f) the number of days to elapse between casting and integrity testing

    (g) preparation of concrete surface of test element for testing using the vibration method

    (h) other particular requirements.

B13.4 Age of test elements at time of testing

In the case of cast-in-place concrete piles, wall elements or barrettes, integrity tests shall not be carried out until the number of days specified in the Project Specification have elapsed since casting.

B13.5 Preparation of element to be tested

Where the method of testing requires the positioning of sensing equipment on a concrete surface at the top of the element to be tested, this surface shall be prepared to expose sound concrete and shall be clean, free from water, laitance, loose concrete, overspilled concrete and blinding concrete, and shall be readily accessible for the purpose of testing.

B13.6 Specialist integrity testing organization

The testing shall be carried out by a specialist organization, subject to demonstration to the Engineer of satisfactory performance on other similar contracts before the commencement of testing.

The Contractor shall submit to the Engineer the name of the specialist integrity testing organization, a description of the test equipment, a test method statement and a programme for executing the specified tests prior to commencement of the Works.

B13.7 Interpretation of tests

The Contractor shall give all available details of the ground conditions, dimensions and construction method of element to be tested, to the specialist company before the commencement of integrity testing in order to facilitate interpretation of the tests. The interpretation of tests shall be carried out by competent and experienced persons.

B13.8 Report

Preliminary results of the tests shall be submitted to the Engineer within 24 hours of carrying out the tests.

The test results and findings shall be reported to the Engineer within 10 days of the completion of each phase of testing.

The report shall identify the test method, include all recorded test data and contain details of the method of interpretation of this data including all assumptions, calibrations, corrections, algorithms and derivations used in the analyses. If the results are presented in a graphical form, the same scales shall be used consistently throughout the report. The units on all scales shall be clearly marked.

B13.9 Anomalous results

In the event that any anomaly is found in the results indicating a possible defect(s) the Contractor shall report such anomalies to the Engineer within 24 hours. The Contractor shall demonstrate to the Engineer that the pile or wall element is satisfactory for its intended use or shall carry out remedial works to make it so in accordance with Clause B1.16. Other than if indicated in the Project Specification, cross-hole sonic logging tubes shall be grouted up after the Contractor has demonstrated that the pile or wall element meets the requirements of the Specification.

B14 Dynamic and rapid load testing of piles

B14.1 General

This section is relevant to the load testing of piles using methods other than static loading.

The design and construction of the load application system shall be satisfactory for the required test. These details shall be submitted to the Engineer prior to the commencement of testing.

B14.2 Project Specification

The following matters are, where appropriate, described in the Project Specification:

(a) type of pile
(b) type of dynamic or rapid load test and method statement of test procedure and the standards to be followed
(c) minimum number of dynamic or rapid load tests to be applied to each pile and procedure to be adopted in testing working piles
(d) special construction detail requirements for test piles
(e) special requirements for pile-testing equipment and arrangement
(f) pile installation criteria
(g) time interval between pile installation and testing
(h) removal of temporary works
(i) details of work to be carried out to the test pile cap or head at the completion of a test
(j) other particular technical requirements.

B14.3 Construction of a preliminary pile to be tested

B14.3.1 Notice of construction

The Contractor shall give the Engineer at least 48 hours notice of the commencement of construction of any preliminary pile which is to be test-loaded.

B14.3.2 Method of construction

Each preliminary test pile shall be constructed in a manner similar to that to be used for the construction of the working piles, and by the use of similar equipment and materials. Any variation will be permitted only with prior approval. Extra reinforcement and concrete of increased strength will be permitted in the shafts of preliminary piles at the discretion of the Engineer.

B14.3.3 Boring or driving record

For each preliminary pile which is to be tested, a detailed record of the conditions experienced during boring, or of the progress during driving, shall be made and submitted to the Engineer within 24 hours. Where the Engineer requires soil samples to be taken or in-situ tests to be made, the Contractor shall present the results without delay.

B14.3.4 Concrete test cubes

For cast-in-situ concrete piles, four test cubes shall be made from the concrete used in the preliminary test pile and from the concrete used for building up a working pile. If a concrete pile is extended or capped for the purpose of testing, a further four cubes shall be made from the corresponding batch of concrete. The cubes shall be made and tested in accordance with BS EN 12390.

The pile test shall not be started until the strength of the cubes taken from any preliminary pile or representative of any working test pile safely exceed the applied stresses at the maximum test load, and shall be not less than twice the average direct stress in any pile section at the maximum test load. The strength of any cubes taken from the pile extension or cap shall also safely exceed the applied stresses at the maximum test load.

B14.4 Preparation of a working pile to be tested
B14.4.1 Notice of construction

The Contractor shall give the Engineer at least 48 hours notice of the commencement of construction of any working pile that is to be load tested.

B14.4.2 Method of construction

Working piles to be tested using dynamic or rapid load testing shall be adequately reinforced to resist the forces applied.

B14.4.3 Preparation of the pile head

The preparation of the pile head for the application of dynamic or rapid test loads shall involve trimming the head, cleaning and building up the pile using materials which will at the time of testing safely withstand the impact stresses. The loading surface shall be flat and at right angles to the pile axis.

Where pile preparation requires drilling holes or welding, this preparation shall not adversely affect the performance of the pile when in service.

Where pile preparation requires the use of a steel plate which is glued to the pile head, this preparation shall not adversely affect the performance of the pile when in service.

B14.5 Safety
B14.5.1 Supervision

The setting-up of pile testing equipment shall be carried out under competent supervision and the equipment shall be checked to ensure that the setting-up is satisfactory before the commencement of load application.

All tests shall be carried out only under the direction of an experienced and competent supervisor conversant with the test equipment and test procedure. All personnel operating the test equipment shall have been trained in its use.

B14.5.2 Safety precautions

Design, erection and dismantling of the pile test reaction system and the application of load shall be carried out according to the requirements of the various applicable statutory regulations concerned with lifting and handling heavy equipment and shall safeguard operatives and others who may, from time to time, be in the vicinity of a test from all foreseeable hazards.

B14.6 Measuring instruments

Calibration certificates valid for the time of pile testing shall be provided to the Engineer for all instruments and monitoring equipment before testing commences.

B14.7 Hammer or propellant

The hammer or propellant, and all other equipment used, shall be capable of delivering an impact or dynamic force sufficient to mobilize the equivalent specified test load.

B14.8 Time of testing

The time between the completion of installation and testing for a preformed pile shall be at least 12 hours or as stated in the Project Specification.

The time between the completion of installation and testing for a cast in place pile shall be at least four days, or as stated in the Project Specification, and shall be such that the pile is not damaged under the stresses during testing.

Preformed piles may be tested during installation when information on hammer performance, soil resistance at time of driving, pile integrity or pile driving stresses is required.

The Contractor shall give the Engineer 48 hours notice of the commencement of the test. Should the Engineer choose not to respond within this time, the Engineer is deemed to have accepted the piles chosen for test as representative of the piles placed.

B14.9 Interpretation of tests

The Contractor shall give all available details of the ground conditions, pile dimensions and construction method to the specialist contractor carrying out the testing in order to facilitate the interpretation of tests.

A competent and experienced person shall carry out the interpretation of the tests.

B14.10 Results

B14.10.1 Report

Preliminary results from the test shall be submitted to the Engineer within 24 hours of the completion of the test.

A full report shall be given to the Engineer within 10 days of the completion of testing.

The report shall identify the test method, include the test data and contain a summary of the method of interpretation of this data including all assumptions, calibrations, corrections, algorithms and derivations used in the analyses. If the results are presented in a graphical form, the same scales shall be used consistently throughout the report. The units on all scales shall be clearly marked.

The report shall contain the following minimum information:

(a) the type of dynamic or rapid load test employed
(b) date of pile installation
(c) date of test
(d) pile identification number and location
(e) length of pile below commencing surface
(f) total pile length, including projection above commencing surface at time of test
(g) length of pile from instrumentation position to toe
(h) hammer type and drop or propellant and calculated force and other relevant details
(i) all recorded test data
(j) blow or load cycle selected for analysis
(k) dynamic or rapid test load achieved
(l) measured pile head movement after each blow or load cycle
(m) permanent residual movement of pile head after each blow or load cycle.

B14.10.2 Additional information

For all piles tested, the following information shall be provided to the Engineer upon request:

(a) maximum measured force applied to the pile head
(b) maximum measured pile head velocity
(c) maximum energy imparted to the pile
(d) permanent set per blow or load cycle
(e) temporary compression
(f) dynamic soil resistance mobilized during the blow
(g) location of possible pile damage.

B14.10.3 Analysis

For piles selected by the Engineer, an analysis of measurements from selected blows or load cycles shall be carried out using a numerical model of the pile and soil to provide the following information:

(a) deduced static load deflection behaviour of the pile head
(b) interpretation and analysis techniques used
(c) parameters used and assumptions made in the analysis
(d) limitations of the method.

B15 Static load testing of piles
B15.1 General

This section is for the testing of a pile by the controlled application of an axial load at the pile head only. It covers vertical and raking piles tested in compression (i.e. subjected to loads or forces in a direction such as would cause the piles to penetrate further into the ground) and vertical or raking piles tested in tension (i.e. subjected to forces in a direction such as would cause the piles to be extracted from the ground). In this section the term 'pile' means either pile or barrette.

The design and full details of the proposed load application, measurement and reaction system shall be satisfactory for the required test. These details shall be submitted to the Engineer at least one week prior to the construction of any test pile or reaction system.

The Contractor shall give the Engineer at least 48 hours notice of the commencement of the test. No load shall be applied to the test pile before the commencement of the specified test procedure.

B15.2 Project Specification

The following matters are, where appropriate, described in the Project Specification:

(a) type of pile

(b) type of test

(c) loads to be applied and procedure to be adopted in testing preliminary piles, including maximum reaction capacity

(d) loads to be applied and procedure to be adopted in proof-testing of working piles, including maximum reaction capacity

(e) whether test is to be compression or tension and number of loading cycles

(f) special materials to be used in construction of preliminary test piles where appropriate

(g) special construction detail requirements for test piles, including requirements for additional reinforcement, increased concrete strength, sampling or in-situ testing

(h) special requirements for pile-testing equipment and arrangement, including requirements for any pile instrumentation

(i) pile installation criteria

(j) time interval between pile installation and testing

(k) removal of temporary works

(l) whether interpretation is required and extent of interpretation

(m) additional records or information required from the load test

(n) displacement transducer stem travel

(o) cut-off level for test

(p) details of work to be carried out to the test pile cap or head at the completion of a test

(q) special requirements for the application of a lateral load to a pile detailed in accordance with the expected conditions of loading (the principles of loading and other relevant details may be adapted from the Specification clauses which follow)

(r) other particular technical requirements.

B15.3 Construction of a preliminary pile to be tested
B15.3.1 Notice of construction

The Contractor shall give the Engineer at least 48 hours notice of the commencement of construction of any preliminary pile that is to be test-loaded.

B15.3.2 Method of construction

Each preliminary pile shall be constructed in a manner similar to that to be used for the construction of the working piles using similar equipment and materials. Additional reinforcement and concrete of

---

increased strength shall be permitted in the shafts of preliminary piles unless specifically precluded in the Project Specification. The Contractor shall inform the Engineer of any differences in the proposed concrete mix or reinforcement. Any variation in method or materials that could affect the pile performance or capacity shall only be permitted with prior approval.

B15.3.3 Boring or driving record

For each preliminary pile, a detailed record of the conditions experienced during boring, or of the progress during driving, shall be made and submitted to the Engineer within 24 hours. Where the Engineer requires soil samples to be taken or in-situ tests to be made, the Contractor shall present the results to the Engineer within 24 hours.

B15.4 Concrete test cubes

For cast-in-situ concrete piles, four test cubes shall be made from the concrete used in any preliminary pile. If a preliminary or working concrete pile is extended or capped after construction, for the purpose of testing, a further four test cubes shall be made from the corresponding batch of concrete. The cubes shall be made and tested in accordance with BS EN 12390.

The pile test shall not be started until the strength of the cubes taken from any preliminary pile or representative of any working test pile safely exceed the applied stresses at the maximum test load, and shall be not less than twice the average direct stress in any pile section at the maximum test load. The strength of any cubes taken from the pile extension or cap shall also safely exceed the applied stresses at the maximum test load.

B15.5 Preparation of a working pile to be tested

If a test is required on a working pile, the Contractor shall cut down or otherwise prepare the pile for testing as required by the Engineer in accordance with Clauses B15.9.1 and B15.9.3.

B15.6 Cut-off level

The cut-off level for a test pile shall be as specified in the Project Specification.

Where the cut-off level of a test pile is below the ground level at the time of pile installation and where it is required to carry out a proof test from that installation level, either allowance shall be made in the determination of the design verification load for friction which may be developed between the cut-off level and the existing ground level, or the pile may be sleeved appropriately or otherwise protected to reduce substantially or eliminate friction which develops over the extended length.

B15.7 Supervision of test

The setting-up of pile testing equipment shall be carried out under competent supervision and the equipment shall be checked to ensure that the setting-up is satisfactory before the commencement of load application.

All tests shall be carried out only under the direction of an experienced and competent person conversant with the test equipment and test procedure. All personnel operating the test equipment shall have been trained in its use.

B15.8 Safety precautions
B15.8.1 General

Design, erection and dismantling of the pile test reaction system and the application of load shall be carried out according to the requirements of current statutory regulations to safeguard everyone, including the public, from all foreseeable hazards.

B15.8.2 Kentledge

Where kentledge is used, the Contractor shall construct the foundations for the kentledge and any cribwork, beams or other supporting structure in such a manner that there will not be differential settlement, bending or deflection of an amount that constitutes a hazard to safety or impairs the efficiency of the operation. The kentledge shall be stacked in such a way to prevent it becoming unstable.

B15.8.3 Tension piles, reaction piles and ground anchorages

Where tension piles, reaction piles or ground anchorages are used to provide the necessary load reaction, they shall be so designed that they will resist the forces applied to them safely and without excessive deformation which could cause a safety hazard during the work. Such piles or anchorages shall be placed in the specified positions, and bars, tendons or links shall be aligned to give a stable reaction in the direction required.

B15.8.4 Testing equipment

In all cases the Contractor shall ensure that when the hydraulic jack and load-measuring device are mounted on the pile head, the whole system will be stable up to the maximum load to be applied.

If, in the course of carrying out a test, any unforeseen occurrence should take place, further loading shall not be applied until a proper engineering assessment of the condition has been made. The engineering assessment shall include a method statement for the safe completion of the test. Data gathering shall continue, if possible and safe, to extract as much data as possible to help diagnose the cause of the unforeseen occurrence.

All hydraulic and other devices shall be operated within their rated capacities, including a safety margin. No leakage from any hydraulic components is acceptable.

The maximum test load expressed as a reading on the gauge in use shall be displayed and all operators shall be made aware of this limit.

B15.9 Reaction systems
B15.9.1 Pile head for compression test

For a pile that is tested in compression, the pile head or cap shall be formed to give a smooth plane surface which is normal to the axis of the pile, sufficiently large to accommodate the loading and settlement measuring equipment, and adequately reinforced or protected to prevent damage from the concentrated application of load from the loading equipment.

Any pile cap shall be concentric with the test pile; the joint between the cap and the pile shall have a strength at least equivalent to that of the pile.

Sufficient clear space shall be made under any part of the cap projecting beyond the section of the pile so that, at the maximum expected settlement, load is not transmitted to the ground by the cap.

B15.9.2 Compression tests

Compression tests shall be carried out using kentledge, tension piles or specially constructed anchorages. Kentledge shall not be used for tests on raking piles except where the test arrangement has been specifically designed for such use and has been approved by the Engineer.

Where kentledge is to be used, it shall be supported on cribwork and positioned so that the centre of gravity of the load is as close as possible to the axis of the pile. The bearing pressure under supporting cribs shall be such as to ensure stability of the kentledge stack.

The weight of kentledge, where known, for each test shall be at least 20% greater than the maximum test load for that test, otherwise 30% if the weight is estimated from the known density and volume of the constituent materials. Additional kentledge required shall be determined taking into account the accuracy of positioning of the centre of gravity of the stack.

The reaction frame support system shall be designed in advance of the testing works. Justification of the capacity of the apparatus shall be provided in advance of the testing works upon request.

Any welding employed to extend or to fix anchorages to a reaction frame shall be carried out so that the full strength of the system is adequate and unimpaired.

B15.9.3 Tension tests

Tension tests may be carried out using compression piles, rafts or grillages constructed on the ground to provide the necessary reaction. In all cases the resultant force of the reaction system shall be coaxial with the test pile.

Where inclined piles or reactions are proposed, full details shall be submitted for approval one week prior to the commencement of testing.

For a pile that is tested in tension, means shall be provided for transmitting the test load axially without inducing moment in the pile. The connection between the pile and the loading equipment shall be constructed in such a manner as to provide a strength at least equal to the maximum load which is to be applied to the pile during the test, with an appropriate factor of safety on the structural design.

Any welding employed to extend or to fix anchorages to a reaction frame shall be carried out so that the full strength of the system is adequate and unimpaired.

B15.9.4 Working piles

If the Contractor plans to use working piles as reaction piles, he shall notify the Engineer of his intention prior to commencement of the Works. Working reaction piles shall not uplift by more than half their specified permissible settlement at working load. The integrity of all working piles used as reaction piles shall be checked within 48 hours of completion of testing, and the integrity test results then provided to the Engineer with 24 hours.

Where working piles are used as reaction piles their movement shall be measured and recorded to within an accuracy of 0.5 mm.

B15.9.5 Spacing

Where kentledge is used for loading vertical piles in compression, the distance from the edge of the test pile to the nearest part of the crib supporting the kentledge stack in contact with the ground shall be such that there is no significant interaction, and shall not be less than 1.3 m.

The centre-to-centre spacing of vertical reaction piles, including working piles used as reaction piles, from a test pile shall be such that there is no significant interaction, and shall not be less than three times the largest diameter of the test pile or the reaction piles, or 2 m, whichever is the greater. Where a test pile has an enlarged base, the same criteria shall apply with regard to the pile shaft, with the additional requirement that no surface of a reaction pile shall be closer to the enlarged base of the test pile than one half of the enlarged base diameter. Where vertical reaction piles penetrate deeper than the test pile, the centre-to-centre spacing of the reaction piles from the test pile shall not be less than five times the diameter of the test pile or the reaction piles, whichever is the greater, unless the base capacity of the test pile is less than 20% of the total ultimate capacity of the test pile.

In cases where reaction piles are not used and where ground anchorages are used to provide a test reaction for loading in compression, no section of fixed anchor length transferring load to the ground shall significantly interact with the test pile, and the centre-to-centre spacing shall be no closer than three times the diameter of the test pile or 2 m, whichever is the greater. Where a test pile has an enlarged base, the same criteria shall apply with regard to the pile shaft, with the additional requirement that no surface of a fixed anchor transferring load shall be closer to the enlarged base of the test pile than a distance equal to the enlarged base diameter.

B15.9.6 Adequate reaction

The reaction frame support system shall be adequate to transmit the maximum test load in a safe manner without excessive movement or influence on the test pile.

B15.9.7 Protection of piles

The method employed in the installation and removal of the reaction system and any other associated Temporary Works shall be such as to prevent damage to any test pile or working pile.

B15.10 Application of load
B15.10.1 Equipment for applying load

The equipment used for applying load shall consist of a hydraulic ram or jack. The jack shall be arranged in conjunction with the reaction system to deliver an axial load to the test pile. Proposals to use more than one ram or jack will be subject to approval by the Engineer of the detailed arrangement. The complete system shall be capable of safely transferring the maximum load required for the test. If the Contractor has concerns that the maximum test load may overstress the pile head, the testing proposal shall include details of preventative measures. The length of stroke of a ram shall be sufficient to cater for deflection of the reaction system under load plus a deflection of the pile head up to 15% of the pile shaft diameter unless otherwise specified or agreed prior to commencement of test loading.

B15.10.2 Measurement of load

A primary calibrated load measuring device shall be used and, in addition, a secondary calibrated pressure gauge shall be included in the hydraulic system. Readings of both the load measuring device and the pressure gauge shall be recorded. In interpreting the test data, the values given by the load measuring device shall normally be used with the pressure gauge readings used as a check for gross error.

The primary load measuring device may consist of a load measuring column, pressure cell or other appropriate system. A spherical seating of appropriate size shall be used to avoid eccentric loading. Care shall be taken to avoid any risk of buckling of the load application and measuring system. Load measuring and application devices shall be short in axial length in order to secure stability. The Contractor shall ensure that axial loading is maintained.

The primary load measuring device shall be calibrated before and after each series of tests, whenever adjustments are made to the device or at intervals appropriate to the type of equipment. The secondary pressure gauge and hydraulic jack shall be calibrated together. Certificates of calibration shall be supplied to the Engineer 48 hours before the commencement of testing.

A record of the load being measured by the secondary pressure guage shall be made at loads of 100%DVL, 100%DVL + 50%SWL, and at the maximum load of each loading cycle to check that the load being measured correlates with the load being measured by the primary load measuring device, within ±5%.

B15.10.3 Control of loading

The loading equipment shall enable the load to be increased or decreased smoothly or to be held constant at any required value. All increments of load shall be maintained within ±0.5% or ±5 kN, whichever is higher, for loads up to 2000 kN. For loads in excess of

2000 kN, all increments of load shall be maintained within ±0.25% or ±20 kN, whichever is lower.

The Contractor shall ensure that axial loading is maintained at all times during the test. If at any stage there are grounds for suspecting that significant eccentric loading is being applied, the load shall be removed, the fault rectified and the test recommenced. Data from this unplanned loading cycle shall be included in the test results.

B15.11 Measuring pile head movement

For a maintained load test, movement of the pile head shall be measured by two reliable independent systems. These shall be:

- optical levelling method; and
- reference beam and displacement transducers/dial guages.

B15.11.1 Optical levelling method

An optical levelling method by reference to a remote datum shall be used.

Where a level and staff are used, the level and scale of the staff shall be chosen to enable readings to be made to within an accuracy of 0.2 mm. A scale attached to the pile or pile cap may be used instead of a levelling staff. At least two reliable independent datum points shall be established at least 15 m apart. Each datum point shall be so situated as to permit a single setting-up position of the level for all readings.

No datum point shall be located where it can be affected by the test loading or any other operations.

B15.11.2 Reference beams and displacement transducers/dial gauges

An independent reference beam or beams shall be set up to enable measurement of the movement of the pile to be made to the required accuracy. The supports for a beam shall be founded in such a manner, and at such a distance from the test pile and reaction system, that movements of the ground do not cause movement of the reference beam or beams which will affect the accuracy of the test. The supports of the beam or beams shall be at least three test pile diameters from the centre of the test pile, or 2 m from the edge of the test pile, whichever is greater.

Check observations of any movements of the reference beam or beams shall be made and a check shall be made of the movement of the pile head relative to a remote reference datum at the start and end and at maximum load for each loading cycle.

The measurement of pile movement shall be made by four transducers or gauges rigidly mounted on the reference beam or beams, bearing on prepared flat surfaces fixed to the pile cap or head and normal to the pile axis. Alternatively, the transducers or gauges may be fixed to the pile and bear on prepared surfaces on the reference beam or beams. The transducers or gauges shall be placed equidistant from the pile axis and from each other. The transducers or gauges shall enable readings to be made to an accuracy of at least 0.1 mm, a resolution of 0.01 mm, and have a stem travel of at least 25 mm or as specified in the Project Specification. Machined spacer blocks may be used to extend the range of reading.

Equivalent electrical displacement measuring devices may be substituted.

B15.12 Protection of testing equipment
B15.12.1 Protection from weather

Throughout the test period all equipment for measuring load and movement shall be appropriately protected from adverse effects of sun, wind and precipitation. Temperature readings shall be taken at the start, end and at the maximum load of each loading cycle.

Table B15.1 Minimum loading times for a maintained load compression proof test

Load*	Minimum time of holding load for a single-cycle pile test	Minimum time of holding for a multi-cyclic pile test
25% DVL	30 minutes	30 minutes
50% DVL	30 minutes	30 minutes
75% DVL	30 minutes	30 minutes
100% DVL	6 hours	6 hours
75% DVL	n/a	10 minutes
50% DVL	n/a	10 minutes
25% DVL	n/a	10 minutes
0	n/a	1 hour
100% DVL	n/a	1 hour
100% DVL + 25% SWL	1 hour	1 hour
100% DVL + 50% SWL	6 hours	6 hours
100% DVL + 25% SWL	10 minutes	10 minutes
100% DVL	10 minutes	10 minutes
75% DVL	10 minutes	10 minutes
50% DVL	10 minutes	10 minutes
25% DVL	10 minutes	10 minutes
0	1 hour	1 hour

* SWL denotes Specified Working Load; DVL denotes Design Verification Load.

B15.12.2 Prevention of disturbance

Construction activity and persons who are not involved in the testing process shall be kept at a sufficient distance from the test to avoid disturbance to the measuring apparatus. Full records shall be kept of any intermittent unavoidable activity and its effects.

B15.13 Test procedure for maintained load compression test
B15.13.1 General

The loading and unloading shall be carried out in stages as shown in Table B15.1. The type of test, single cycle or multi-cyclic, shall be in accordance with the Project Specification. Any other particular requirements stated in the Project Specification shall be complied with.

Following each application of an increment of load, the load shall be maintained at the specified value for not less than the period shown in Table B15.1 and until one of the following rate of settlement criteria are satisfied:

- For pile head displacements of less than 10 mm, each load increment shall be maintained until the rate of settlement is reducing and is ≤ 0.1 mm/hour.
- For pile head displacements between 10 mm and 24 mm, each load increment shall be maintained until the rate of settlement is reducing and is $\leq 1\% \times$ pile head displacement/hour.
- For pile head displacements of greater than 24 mm, each load increment shall be maintained until the rate of settlement is reducing and is ≤ 0.24 mm/hour.

The rate of settlement shall be measured over a minimum period of 30 minutes.

For any period when the load is constant, time and settlement shall be recorded immediately on reaching the load, and at least at the following rates:

- For the first 15 minutes — every 5 minutes.
- For the first 1 hour — every 15 minutes.
- For the first 4 hours — every 30 minutes.
- Thereafter — every 1 hour.

B15.13.2 Proof load test procedure for working compression piles

The maximum load which shall be applied in a proof test shall be the sum of the Design Verification Load (DVL) plus 50% of the Specified Working Load (SWL).

B15.13.3 Load test procedure for preliminary compression piles

The maximum load which shall be applied in a preliminary load test shall be specified in the Project Specification. The loading and unloading shall be carried out in accordance with Table B15.1 and Clause B15.13.1. Any particular requirements stated in the Project Specification shall be complied with.

Following the completion of the proof load test in accordance with Table B15.1, the load shall then be restored in two stages (DVL, DVL + 50%SWL) with each load step held for at least 1 hour. Then the load shall subsequently be increased by increments of 25%SWL until the maximum specified load for the test is reached. For each application of an increment of load, the load shall be maintained at the specified value in accordance with Clause B15.13.1.

When the load applied to the test pile cannot be maintained within the specified settlement rate criteria, or the pile head velocity has not reduced within any 2 hour period after the application of a load increment, an attempt shall be made to stabilize the load at the previous load increment:

- If the load can be stabilized, the load shall be held for 30 minutes and then increased in increments of 10%SWL, each held for 30 minutes, until a load is reached where the load cannot be stabilized.
- If the load fails to stabilize, the load shall again be reduced to the highest previous load increment at which the load can be stabilized and the load held at that level for 30 minutes, and then increased in increments of 10%SWL until a load is reached where the load cannot be stabilized. For each of these increments, the load shall be held for 30 minutes.

The load shall then be reduced in five approximately equal stages to zero load, with penetration and load at each stage and at zero load being recorded.

B15.13.4 Testing of piles designed to carry load in tension

The testing of piles designed to carry load in tension shall follow the same procedure as specified in Clauses B15.13.1, B15.13.2 and B15.13.3.

B15.14 Presentation of factual results
B15.14.1 Results to be submitted

During the progress of a test, all measurements taken shall be available for inspection by the Engineer.

Results shall be submitted to the Engineer as follows:

(a) A preliminary copy of the test records shall be submitted, unless otherwise specified, within 24 hours of the completion of the test. This shall show for each stage of loading, the period for which the load was held, the load and the maximum pile movement at the end of the stage.
(b) A pile test report including the completed schedule of recorded data as prescribed in Clause B15.14.2 within 10 days of the completion of the test.

B15.14.2 *Schedule of recorded data*

The Contractor shall provide information about the testing in accordance with the following schedule where applicable:

(a) *General*

All types of pile
Site location
Contract identification
Principal contractor
Piling contractor
Testing contractor
Engineer
Client/Employer
Date and time of test

(b) *Pile details*

All types of pile
Identification (number and location)
Ground conditions
Specified Working Load (SWL)
Design Verification Load (DVL)
Commencing surface level at pile position
Head level at which test load was applied
Type of pile
Vertical or raking, compression or tension
Shape and size of cross-section of pile, and position of any change in cross-section
Shoe or base details
Head details
Length in ground
Level of toe
Dimensions of any permanent casing
Whether pile incorporates any instrumentation

Concrete piles
Concrete mix/grade
Aggregate type and source
Cement type and cement replacement and type where used
Admixtures
Slump
Concrete cube test results for pile and cap
Date of casting of a precast pile
Reinforcement

Steel piles
Steel quality
Coating
Filling or core materials — type and quality

(c) *Installation details*

All types of piles
Dates and times of boring/concreting or driving of test pile
Difficulties and delays encountered
Any non-conformances noted during test pile construction
Date and time of casting concrete pile cap
Results of any integrity tests

Bored, CFA or DA piles, or barrettes
Type of equipment used and method of boring or excavating
Temporary casing — diameter, type and length
Record of ground conditions during boring or excavating

Date, time and method of placing concrete
Volume of concrete placed
Use of support fluid and testing records
Base grouting or other relevant records

Driven preformed or cast-in-situ concrete piles or driven steel piles
Method of support of hammer and pile
Driven length of pile at final set
Hammer type, and size or weight
Dolly and packing, type and condition
Driving log (depth, hammer drop, blows per 250 mm, interruptions or breaks in driving)
Set in number of blows to produce final penetration of 25 mm
Redrive check, time interval and set in number of blows to produce penetration of 25 mm or other agreed amount
At final set and at redrive set: for a drop hammer or for a single acting hammer the length of the drop or stroke; for a diesel hammer the length of the stroke and the blows per minute; for a double acting hammer the operating pressure and the number of blows per minute
Condition of pile head after driving
Use of a follower
Use of pre-boring
Use of water jetting
Lengthening of piles
Method of placing concrete

(d) *Test procedure*
All types of pile
Mass of kentledge
Tension pile, ground anchorage or compression pile details
Plan of test arrangement showing position and distances of kentledge supports, rafts, tension or compression piles or ground anchorages, and supports to the pile movement reference system
Jack capacity
Method of load measurement
Method(s) of penetration or uplift measurement
Calibration certificates
Temperature readings

(e) *Test results*
All types of pile
In tabular form: all measurements
In graphical form: load plotted against pile head movement; load plotted against time
Ambient temperature records during test

B15.15 Presentation of test interpretation
B15.15.1 Need for interpretation

If specified by the Project Specification, the Contractor shall present an interpretation of the test results in accordance with Clause B15.15.2 or B15.15.3.

B15.15.2 Working load test interpretation

The Contractor shall make a statement regarding the shape of the test curve up to its maximum point, including the compatibility of the curve shape with the likely mobilization of the assumed design parameters.

The Contractor shall make a statement on the adherence of the results to the specified displacement limits for DVL and for

$DVL + \frac{1}{2}SWL$. Where any result is greater than 80% of either of the limits, the statement shall explain the significance of the results in terms of likely future adherence for the other working piles. Where a limit has been exceeded, the Contractor shall make a proposal in accordance with Clause B1.16.

B15.15.3 Preliminary load test interpretation

The Contractor shall make a statement on the adherence of the results to the specified displacement limits for DVL and for $DVL + \frac{1}{2}SWL$. Where any result is greater than 80% of either of the limits, the comments shall explain the significance of the results in terms of likely future adherence for the other working piles. Where a limit has been exceeded, the Contractor shall make an explanation of the likely reasons for the non-performance with a proposal to the Engineer regarding the design and/or construction of the working piles and/or regarding the requirements of the Project Specification.

Where the use of strain gauges and extensometers, or similar, within the preliminary pile have been specified, the interpretation shall be in two parts: firstly an interpretation using the externally measured loads and displacements, then secondly a more detailed interpretation using the strain gauge and extensometer data. The interpretation of the strain gauge and extensometer data shall convert the raw instrument data into load and displacement respectively. All assumptions shall be clearly stated. Where any of the data appears to be erroneous or inconsistent, then any assumptions shall be stated with reasons as to which pieces of data are correct and can be relied upon, and which are to be discounted.

The interpretation shall separate the shaft and base components, and assess the likely shaft capacities mobilized in each layer.

For a preliminary pile test where the maximum applied load did not reach the ultimate pile capacity, the interpretation shall be based primarily on the measured data, but, if required, shall extend to an extrapolation of the results using the methods proposed by Chin (1970) and/or Fleming (1992), or similar, in order to deduce the likely ultimate pile shaft and base capacities.

The interpretation shall compare the pile shaft and base capacities to the assumed design parameters and comment on their applicability. Reference shall be made to the most appropriate borehole logs and to the pile construction log. If the interpretation concludes that the pile design needs to be revised, then recommendations for the changes needed shall be given together with an indication of the effects on the working piles. If the interpretation concludes that there are any doubts about any working piles already installed, then the Contractor shall make a proposal in accordance with Clause B1.16.

If any changes resulting from the preliminary pile test data have any impact on cost and/or programme, then the Contractor shall inform the Engineer immediately.

B15.16 Completion of a test
B15.16.1 Removal of test equipment

On completion of a test, all measuring equipment and load application devices shall be dismantled and checked. All other Temporary Works, test equipment including kentledge, beams and supporting structures, shall be removed from the test pile location. Measuring and other demountable equipment shall be stored in a safe manner so that it is available for further tests if required, or removed from the Works.

Temporary tension piles and ground anchorages shall be cut off below ground level, and off-cut materials removed. The ground shall be made good to the original commencing surface level.

B15.16.2 Preliminary pile

Unless otherwise specified, the head of each preliminary test pile shall be cut off below ground level, off-cut material shall be removed from the Works and the ground made good to the original commencing surface level.

B15.16.3 Working pile test cap

On completion of a proof test on a working pile, the test pile cap shall be prepared as required by the Project Specification and left in a state ready for incorporation into the structure.

Specification for piling and embedded retaining walls. Thomas Telford, London, 2007

B16 Piles with sleeves and/or coatings
B16.1 General

Where pile sleeves and/or coatings are specified, whether for reducing friction or for providing a protective barrier, but the particular method is not specified, the Contractor shall provide full details of the method to be employed. The Contractor shall ensure that any product used will be compatible with the ground conditions into which it will be installed. Particular requirements are detailed in the Project Specification.

B16.2 Project Specification

The following matters are, where appropriate, described in the Project Specification:

(a) the type and particular description of the sleeving method and/or coating to be used

(b) the purpose of the sleeving or coating

(c) the numbers or other identification of piles to be sleeved or coated

(d) the length of pile to be sleeved or coated

(e) preparatory pre-boring or other work necessary for proper application of the method

(f) depth, diameter and means of ensuring temporary stability of any pre-boring where required

(g) designated manufacturer's name and details where a proprietary product is required

(h) whether any annulus surrounding the sleeving is to be filled with grout or other suitable self-hardening fluid

(i) details of any pile testing or trial piles required to demonstrate the effectiveness of the method

(j) other particular technical requirements.

B16.3 Bituminous or other coating materials
B16.3.1 General

Where a proprietary product is used, the process of cleaning the surface of the sleeving or the pile, and the conditions and methods of application shall conform with the manufacturer's current instructions. All materials shall conform to the manufacturer's specification, which shall be given to the Engineer at least one week before any coating is applied.

B16.3.2 Protection from damage

Where a coating material has been applied to sleeving or to a preformed pile prior to installation, it shall be protected from damage during handling and transportation. In the event of damage to the coating, it shall be made good on site to the same specification as the original coating prior to being installed. Where bituminous materials are involved, precautions shall be taken as necessary in hot weather to prevent excessive flow or displacement of the coating. The coating shall be adequately protected against direct sunlight and, if stacked, the piles shall be separated to prevent their coatings sticking together.

B16.3.3 Installation

Sleeves with applied coatings shall not be placed, and preformed piles with applied coatings shall not be driven, when the air temperature is such that the coating will crack, flake or otherwise be damaged prior to entry into the ground. Where bituminous materials are involved, installation shall be carried out while the temperature is at or above 5°C or as stated in the manufacturer's instructions.

B16.4 Sleeving
B16.4.1 General

Detailed design of the sleeving shall be submitted to the Engineer at least one week prior to installation within the piles. All materials shall conform with the manufacturer's specification, which shall be given to the Engineer at least one week before the material is used.

Where the sleeving is to be made from a series of short sections, it shall be watertight.

The dimensions and quality of sleeving shall be adequate to withstand the stresses caused by handling and installation without damage or distortion.

B16.4.2 Lengthening of steel sleeving

The lengthening of steel sleeves by adding an additional length of steel sleeving during construction shall be carried out in accordance with the relevant procedures in Section B6.

B16.5 Inspection of coatings

The Engineer may call for piles to be partially exposed or extracted. Where significant damage to the coating is found to have occurred, the Contractor shall submit a method statement for the repair or replacement of the coating.

Specification for piling and embedded retaining walls. Thomas Telford, London, 2007

B17 Instrumentation for piles and embedded retaining walls
B17.1 General

This section is relevant to the instrumentation to monitor the behaviour of piles and embedded retaining walls.

The installation, monitoring and reporting shall be carried out by a specialist organization, subject to demonstration to the Engineer of satisfactory performance.

The staff carrying out the installation, monitoring and reporting of the results shall be competent and experienced with the type of instrumentation used.

B17.2 Project Specification

The following matters are, where appropriate, described in the Project Specification:

(*a*) aims and objectives of the instrumentation

(*b*) trigger values of measured parameters that require immediate special attention including reporting to the Engineer

(*c*) the type of instrumentation required

(*d*) the pile or wall element numbers and locations in which the instrumentation is to be installed

(*e*) the depth or location within the pile or wall where the instrumentation is to be installed

(*f*) time at which the base readings should be taken

(*g*) time schedule for readings over the duration of the project

(*h*) maximum time lapse between taking the readings and reporting to the Engineer

(*i*) expected load, pressure, displacement or strain range for which results are required including resolution

(*j*) type of loading, compressive or tensile

(*k*) type of output required

(*l*) monitoring equipment (manual or automatic)

(*m*) whether remote or direct reading is required

(*n*) whether the instrumentation monitoring equipment will become the property of the Employer

(*o*) whether surveying of the pile or element head terminal is required, to what grid and datum, and frequency of surveying

(*p*) responsibility for instrumentation: installation, calibration, monitoring and interpretation of results

(*q*) other special instrumentation

(*r*) other particular technical requirements.

B17.3 Type of instrumentation

Where the installation of instrumentation is called for, the type of instrumentation shall be one of the following, as specified in the Project Specification:

(*a*) extensometer

(*b*) inclinometer

(*c*) load cell

(*d*) pressure cell

(*e*) strain sensor

(*f*) surveying

(*g*) other special instrumentation not covered by the above.

Other methods may be considered by the Engineer subject to satisfactory evidence of performance. All equipment used shall be suitable for its specified purpose.

The instrumentation shall be robust and shall be a proprietary system supplied by a reputable supplier and shall be installed by a specialist organization.

B17.3.1 Extensometers
B17.3.1.1 General

Where the instrumentation is to be installed in a reinforced concrete element, it shall be securely attached to the reinforcement cage so that no component is displaced during placing of the reinforcement cage or concreting. During concreting, the tubing shall be adequately sealed at both ends to prevent the ingress of concrete and, in the case of a magnetic extensometer, it shall then be filled with clean water. Compression or extension couplings shall be installed as necessary for the movement range specified in the Project Specification.

Where the instrumentation is to be installed in a steel element, it shall be securely attached to the element. The attaching system shall be such as to withstand the effects of installation of the element in the ground.

B17.3.1.2 Rod extensometers

The system will be such that the anchor and reference blocks shall be securely attached to the pile or wall element. The rod shall be made so that it will not corrode or distort nor change its length due to heat or water changes. The extensometer shall be entirely free to move and unrestricted by its sleeving.

The top of the rod shall incorporate a range adjuster with a travel of 25 mm.

If direct measurements are specified, the displacement measuring device shall consist of a suitable mechanical (e.g. analogue dial gauge) or electrical (electronic device with calibrated digital readout) displacement transducer.

The orientation of the device relative to the top of the tube shall be constant for every reading. If remote reading is specified, the readings shall be made by a linear potentiometer or other suitable device, securely held in place.

B17.3.1.3 Magnetic extensometers

The system shall be such that the magnets are securely cast into the pile concrete. The internal diameter of the tube shall be compatible with the proposed measuring device to avoid variations in readings due to non-concentric position of the reading device in the tube. The tube shall be kept full of water during monitoring and the temperature of the water shall be monitored and recorded to an accuracy of $1°C$.

One of the following devices shall be used for measuring depths according to the accuracy required:

(a) Reed switch mounted on a hand-held tape similar to that used for ground instrumentation. Readings shall be obtained by sighting a precise level on to the tape at the pile head to achieve a repeatable accuracy of better than 0.3 mm. The absolute level of the precise level shall be monitored during the test; or

(b) Reed switch mounted on a cable and attached to a micrometer device at the head of the pile such that readings can be established to an accuracy of better than 0.1 mm. The absolute level of the micrometer device shall be monitored during the test.

B17.3.2 Inclinometers
B17.3.2.1 General

The system will be such as to comprise an access tube with orthogonal oriented keyways to provide access for the measurement instrumentation. The system shall consist of corrosion-proof access tubing which shall have four longitudinal internal keyways on two orthogonal axes. The keyways shall be continuous robust and so formed over the total length of the tubing to allow the inclinometer to be successfully deployed. The 'azimuth deviation' of the keyways shall be recorded for each inclinometer access tube and shall be taken into account when processing the data. Where azimuth correction is necessary, the Contractor shall submit details of the method used. Base readings shall be taken to establish the profile of the access tube immediately after installation. The base readings shall be used to establish the datum profile and shall be notified to the Engineer.

The access tubing shall be securely attached to the element. In the particular case of installation in a cast-in-situ element, the access tube shall be attached to the reinforcement cage so that no component is displaced during placing of the reinforcement cage or concreting. During concreting, the tubing shall be adequately covered at both ends to prevent the ingress of concrete. Alternatively, nominal 100 mm internal diameter steel duct, sealed at the lower end and fitted with a removable screw cap at the upper end, shall be attached to the reinforcement cage. The inclinometer access tube can then be installed afterwards using a cementitious grout containing a non-shrink additive.

Inclinometers shall give a stable output over the temperature and inclinations expected and be calibrated with the in-service cables and readout units.

B17.3.2.2 Torpedo type

The inclinometer readout equipment shall comprise a biaxial torpedo complete with operating cable on a reel, readout unit or data logger. The biaxial torpedo and operating cable shall be fully waterproof. The torpedo shall have a gauge length of 0.5 m and be capable of negotiating a tube curvature of 3 m. The cable should be of an appropriate length with markings at 0.5 m intervals. The readout unit or data logger shall have an alphanumeric display . The Contractor shall supply a calibration frame for checking the readout equipment over the expected range of movements.

Readings shall be taken for the full length of each inclinometer access tube for faces A, B, C and D. The top wheel of the torpedo will travel up face A first followed by faces B, C and D.

If a full set of readings cannot be obtained due to accumulation of debris in the access tubes, the Contractor shall flush the tubes with water to remove the debris until readings can be taken.

The readings shall be taken in increments of 0.5 m starting from the base of the access tube. All readings shall be recorded and reported to the nearest 0.1 mm. The readings are considered to be sufficiently accurate only if the 'face errors' in the A/B plane and the C/D plane are less than 1.5 mm.

The data shall be processed assuming a fixed base unless otherwise agreed with the Engineer. The surveyed cumulative horizontal movement of the access tube shall be reported to an accuracy of 0.5 mm. The data shall also be processed assuming a specified offset at the top of the access tube at the request of the Engineer.

The processed data for faces A, B, C and D shall be tabulated to show the following:

(*a*) deviations and face errors
(*b*) mean deviation
(*c*) change in mean deviation
(*d*) cumulative mean deviation
(*e*) displacement profiles.

Items (*c*) and (*e*) should also be presented graphically to show the deviation or displacement plotted against depth.

B17.3.2.3 Non-torpedo types

Non-torpedo types are those where sensors are installed in the access tube and read remotely. The sensors shall be incorporated such that their effective gauge length is 0.5 m and each gauge length jointed on to the next gauge length to provide an accurate continuous deformation profile in orthogonal planes to a cumulative horizontal movement of the access tube to an accuracy of 0.5 mm.

The remote reading equipment shall be located and protected such that data can be logged and recorded throughout the life of the project.

B17.3.3 Load cells

The loads to be measured may be either compressive or tensile, as specified in the Project Specification. The load cells shall have a response time of 2 seconds or less in response to monotonically increasing or decreasing loading.

The monitoring system shall provide a stable signal and any temperature or cable length or other effects on the load signal shall be included in the cell calibration and permissible tolerance. If a hydraulic or pneumatic system is used it shall be rated for, and have been tested to, twice the anticipated maximum pressure. The load cells and cabling shall be compatible with their intended position within a pile and shall be unaffected by the presence of water or fluid concrete. They shall be sufficiently strong to withstand additional loading due to placing of the reinforcement cage and concreting.

As the cell will give a direct reading of load, it should be positioned so that all the load passes through it without any eccentricity or the possibility of arching that would reduce the recorded load on the cell. The loading system shall be safe and stable.

B17.3.4 Pressure cells

The pressure cells shall have a response time of 2 seconds or less in response to monotonically increasing or decreasing loading.

The monitoring system shall provide a stable signal and any temperature or cable length or other effects on the pressure signal shall be included in the cell calibration and permissible tolerance. If a hydraulic, oil, pneumatic or mercury-based system is used it shall be rated for, and have been tested to, twice the anticipated maximum pressure. The pressure cells and cabling shall be compatible with their intended position within a pile and shall be unaffected by the presence of water or fluid concrete. They shall be sufficiently strong to withstand additional loading due to placing of the reinforcement cage and concreting.

If the pressure cell is to be placed against hardened concrete on one or both sides, it shall be provided with a re-pressurizing tube to ensure that no gap occurs between the hardened concrete and the pressure cell.

B17.3.5 Strain sensors
B17.3.5.1 General

The strain sensors shall have a response time of 10 seconds or less. The monitoring systems shall provide a stable signal and any temperature or cable length or other effects on the signal shall be included in the permissible tolerance. The strain sensors and cabling shall be compatible with their intended position within a pile and shall be unaffected by the presence of water or fluid concrete. They shall be sufficiently strong to withstand additional loading due to placing of the reinforcing cage and concreting or pile driving.

B17.3.5.2 Strain sensors attached to steel or pre-cast concrete piles

The strain sensors shall be securely attached to a mounting plate welded to the steel for steel piles, or securely bolted to the reinforced concrete for pre-cast concrete piles. The welding procedures of Section B6 shall be followed. The method of attachment shall be sufficiently strong so the sensor is not displaced, nor the wires or cables damaged by driving.

B17.3.5.3 *Strain sensors embedded in concrete*	Strain sensors for cast-in-situ piles shall be either the embedment variety or shall be securely attached to mounting plates welded to 'sister' bars which shall be attached to the reinforcement cage. The methods of attachment shall be such that the sensor is not displaced, nor the wires or cables damaged by placing of the reinforcement, concreting, driving or other processes.

The attachment of sister bars shall be at two locations only, so that the amount of bending induced in the bars and measured by the strain sensors changes by less than 50 microstrain.

B17.3.6 *Surveying*	The surveying system shall consist of an optical instrument that can be used to either manually or automatically sight on to a target. The system will be either:

(*a*) precise levelling to measure level differences; or
(*b*) geodetic surveying to measure the spatial juxtaposition of the instrument and targets.

B17.3.6.1 *Precise levelling*	The precise levelling measurements shall be made in relation to a datum as agreed by the Engineer. The vertical stability with time of the datum must be established before the project starts and be accepted by the Engineer. The precision of the levelling shall be agreed by the Engineer. The targets shall be securely fixed to the element to be surveyed or be of the demountable type with a locking system that gives a repeatable positioning of the target with a variation in level that is less than the resolution of the surveying system. The levelling instrument shall be equidistant from change points. The precise levelling circuit shall start and finish on the datum and the readings, calculations and closing error reported.

B17.3.6.2 *Geodetic surveying*	The locations of the instrument and targets shall be agreed by the Engineer with at least two fixed datums to which the targets shall be related. The spatial movement of the datums must be established and agreed by the Engineer before the project begins. The precision of the survey must be agreed by the Engineer.

The manual or automatic readings taken at the instrument will be such that a vector can be calculated to relate the optical instrument and the target in three dimensions related to the distance between the recording. The locations of the instrument shall be fixed with a device to lock the instrument in the same position for every set of readings. The targets used shall be securely fixed to the element being surveyed or be of the demountable type with a locking system that gives a repeatable positioning of the target. The repeatable positioning systems for the instrument and targets shall be such as to give an error in the spatial position of less than the resolution of the surveying system.

B17.3.7 *Other Special Instrumentation*	If other Special Instrumentation is required, it shall be included in the Project Specification. The Special Instrumentation shall be fully described in its action, durability, repeatability and the effects of ambient conditions. The specification of the Special Instrumentation shall be fully described with satisfactory evidence of performance in the likely project conditions.

B17.4 Installation	Except where explicitly specified in the Project Specification, no instrumentation device, tubing or cable shall be placed in the concrete cover zone.

B17.4.1 Personnel	The Contractor shall submit details of the supplier of the instrumentation and curricula vitae for the staff from any specialist organization(s) involved in the installation and/or monitoring of the instrumentation on site and/or analysis of the readings. The Contractor shall also submit details of projects where the specialist organization(s) has successfully installed and monitored the specified type of instrumentation.
B17.4.2 Protection	Terminal boxes at the head of the pile shall be protected by a lockable robust steel cap. The pile head shall then be fenced off with clearly visible barriers which shall be maintained for the duration of the monitoring programme.
	Any cable running along the ground shall be clearly marked and adequately protected to prevent the cable posing a safety hazard or becoming damaged.
B17.4.3 Surveying	The pile or element head terminal (top of access tubing or as specified in the Project Specification) shall be surveyed, if specified in the Project Specification. The level will be determined to an accuracy of 1 mm relative to the datum specified in the Project Specification. The grid coordinates will be determined to an accuracy of 2 mm relative to the grid specified in the Project Specification. Other measurements shall be to the specified accuracy.
B17.5 Monitoring	The instrumentation monitoring schedule shall be coordinated with the construction activity schedule, maintenance schedule and operations as defined in the Project Specification.
B17.5.1 Monitoring equipment	Monitoring shall be carried out either directly at the element, or remotely from a monitoring cabin as specified in the Project Specification.
	The monitoring equipment shall be appropriate for the situation in which it is to be used. The manufacturer's recommendations for its use shall be followed. Where it has not been possible to follow the manufacturer's recommendations, the fact shall be reported together with the reason and details of the alternative procedure.
	The monitoring equipment shall remain on site for the duration of the monitoring programme, except when necessary for it to be calibrated, after which it shall be returned to site. The same monitoring equipment shall be used at each position. Where this is not possible, the fact shall be reported together with the reason and a calibration of the replacement equipment be made against the original equipment.
B17.5.2 Readings	Readings shall be reported in terms of the basic property measured (e.g. volts, hertz, mm of mercury) and shall be converted to SI or derivative units by means of a calibration constant. The calibration constant and the range over which it is applicable shall be clearly stated. A calibration certificate shall be provided for each instrument.
	Readings shall be taken at the times or time intervals specified in the Project Specification.
	The readings can be taken by a manual or automatic system but must be relayed to the Engineer in the form required within the time lapse in the Project Specification.
B17.5.3 Calibration and data checking	The instrumentation shall be calibrated prior to incorporation into the Works and a certificate of calibration shall be submitted to the

Engineer. For strain sensors a calibration constant can be provided if evidence is available showing that the variation in calibration constant will not vary maximum readings outside the specified tolerance.

The instrumentation shall be calibrated so that its behaviour has been monitored over the range specified in the Project Specification. The Contractor shall demonstrate that the type of instrumentation can provide a stable, reproducible and repeatable calibration within the likely temperature range the instrument will experience.

All data shall be checked by the Contractor for errors prior to submission. If erroneous data are discovered (e.g. face errors for inclinometer readings greater than 1.5 mm), the Contractor shall take a second set of readings immediately. If the errors are repeated, the Contractor shall determine the cause of the error. Both sets of readings shall be processed and submitted, together with the reasons for the errors and details of remedial action, if any. The Contractor shall rectify any faults found in the instrumentation system for the duration of the specified monitoring period.

All computer data files and calculation sheets used in processing the data shall be preserved until the end of the contract. They shall be made available for inspection at the request of the Engineer.

B17.6 Report

The results shall be given to the Engineer In accordance with the Project Specification.

The report shall contain the following:

(*a*) the date and time of each reading
(*b*) the weather
(*c*) the name of the person who made the reading on site and the name of the person who analysed the readings together with their company affiliations
(*d*) the element reference number and the depth and identity number of the instrumentation
(*e*) any damage to the instrumentation or difficulties in reading
(*f*) the condition of the element and related construction (e.g. if the reading is being made during a load test, the stage of the test; if a wall is being excavated alongside, the depth of the excavation, etc.)
(*g*) the calibration constants and/or equations that are being applied and the dates they were determined
(*h*) a table comparing the specified results with any previous readings and with the base readings
(*i*) a graph showing variation of load or pressure or vertical movement or strain with time or horizontal movement with depth. Key dates should be marked with a brief explanation of their significance.

Columns of numbers shall be clearly labelled together with units. Numbers should not be reported to a greater accuracy than is appropriate. Graph axes should be linear and clearly labelled together with units.

B18 Support fluid
B18.1 General

Where a support fluid is used for maintaining the stability of an excavation the properties and use of the fluid shall be such that the following requirements are achieved:

(*a*) continuous support of the excavation
(*b*) solid particles are kept in suspension (except in the case of water and some polymers)
(*c*) the fluid can be easily displaced during concreting
(*d*) the fluid does not coat the reinforcement to such an extent that the bond between the concrete and reinforcement is impaired
(*e*) the fluid shall not cause pollution of the ground and groundwater before, during or after use
(*f*) the fluid and/or its constituents shall not possess properties which are potentially harmful to site operatives without reasonable precautions.

Details of the type of support fluid, manufacturer's instructions and certificates for the constituents and mix proportions, including all additives, shall be submitted by the Contractor at the time of tender.

B18.2 Project Specification

The following matters are, where appropriate, described in the Project Specification:

(*a*) minimum material testing requirements and schedule of testing
(*b*) environmental restrictions on use
(*c*) testing of water and schedule of testing
(*d*) submission of testing records to the Engineer
(*e*) other particular technical requirements.

B18.3 Evidence of suitability of support fluid

The Contractor shall provide details of the use of, and testing schedule of, the support fluid to demonstrate that it will meet the specified requirements. These details shall be submitted to the Engineer at least 14 days prior to the commencement of the Works and shall include:

(i) Evidence of previous experience with this support fluid and justification for its suitability for these ground conditions and method of construction; particular issues which shall be addressed are the types and sources of the support fluid constituents, time of construction of the pile, panel or barrette, ambient temperature, soil and groundwater chemistry.
(ii) Results of representative laboratory or field mixing trials with the support fluid to demonstrate compliance with the Specification.
(iii) Details of the tests to be used for monitoring the support fluid during the Works and the compliance values for these tests, presented in the form of Table B18.1, or similar.

B18.4 Materials
B18.4.1 Water

If water for the Works is not available from a public supply, the Contractor shall use an alternative source that shall comply with the requirements of BS EN 1008.

When required by the Project Specification, the Contractor shall arrange for tests of the water for the Works to be carried out in accordance with the specified schedule before and during the progress of the work. The frequency of testing shall be as stated in the Project Specification.

B18.4.2 Additives to the water

All solid additives shall be stored in separate waterproof stores with a raised floor or in waterproof silos which shall not allow the material to become contaminated.

Table B18.1 Tests and compliance values for support fluid prepared from bentonite

Property to be measured	Test method and apparatus	Compliance values measured at 20°C		
		Freshly mixed	Ready for re-use	Sample from excavation prior concreting
Density	Mud balance			
Fluid loss (30 minute test)	Low-temperature test fluid loss			
Filter cake thickness	Low-temperature test fluid loss			
Viscosity	Marsh cone			
Shear strength (10 minute gel strength)	Fann viscometer			
Sand content	Sand screen set			
pH	Electrical pH meter to BS 3445; range pH 7 to 14			

Additives shall generally be used in accordance with the manufacturer's instructions unless suggested variations can be demonstrated to be fit for proposed purpose.

If used, bentonite shall be of a quality in accordance with Publication 163 (1988) of the Engineering Equipment and Materials Users Association.

B18.5 Mixing of support fluid

The constituents of the fluid shall be mixed thoroughly to produce a homogeneous mix. The temperature of the water used in mixing, and of the support fluid at the time of commencing concrete placement, shall not be less than 5°C.

B18.6 Compliance testing of support fluid

The Contractor shall carry out testing of the support fluid in accordance with the Contractor's schedule and the Project Specification to demonstrate compliance within the limits for each test. The Contractor shall establish a suitably equipped and properly maintained site laboratory for this purpose and provide skilled staff and all necessary apparatus to undertake the sampling and testing.

In accordance with the schedule of testing, freshly prepared or reconditioned support fluid shall be proven by sampling and testing to be within the compliance limits and the results recorded. These records shall be submitted to the Engineer as required by the Project Specification.

Details of the method, frequency and locations for sampling and testing support fluid from the excavation shall be submitted by the Contractor at least 14 days prior to the commencement of the Works. At least one sample immediately prior to placing reinforcement and concrete shall be taken and tested from the base of the excavation and one near to the top.

If tests show that the support fluid does not comply with the Specification, it shall be replaced before concreting the pile. However, if following suitable reconditioning and/or partial replacement the support fluid does then comply with the Specification, it may remain in the excavation.

B18.7 Spillage and disposal of support fluid

All reasonable steps shall be taken to prevent the spillage of support fluid in areas outside the immediate vicinity of excavation. Discarded fluid shall be removed from the Works without undue delay. Any disposal of fluid shall comply with the requirements of current legislation and all relevant authorities.

B19 General requirements for concrete and steel reinforcement
B19.1 General

All materials shall be in accordance with Section B1 and B19 of this Specification, BS EN 1536, BS 8500 and the Project Specification, except where there may be conflict of requirements, in which case the most onerous Specification shall take precedence.

Steel reinforcement shall be in accordance with BS 4449. It shall be manufactured and supplied to a recognized third party product certification scheme meeting the requirements of Clause 8.2 of BS 4449, or shall be subject to acceptance testing on each batch in accordance with Annex B of BS 4449.

In this section the term pile shall refer to piles, diaphragm wall panels and barrettes.

Requirements for the Project Specification shall be stated in Table B19.4.

B19.2 Concrete
B19.2.1 Strength class

Strength classes of concrete shall be in accordance with BS 8500 and shall be denoted by the characteristic 28 day test cylinder strength followed by the characteristic 28 day test cube strength in N/mm^2 (e.g. C28/35).

B19.2.2 Composition of concrete

The composition of concrete with respect to materials and limiting mix proportions shall be in accordance with the requirements of BS 8500 for designed concrete or prescribed concrete.

The total chloride content of the concrete shall be calculated from the mix proportions and the measured chloride content of each of the constituents in accordance with Clause 5.3 of BS 8500-2 and shall not exceed the limits given in Table B19.1, in accordance with Clause A.9 of BS 8500-1. The limit for pre-stressed concrete shall also apply to the structural concrete of post-tensioned pre-stressed concrete work unless there is an impermeable and durable barrier, in addition to any grout, between the main concrete and the tendons.

The cement content in any structural concrete shall be not less than $300 \, kg/m^3$. Where concrete is to be placed under water or support fluid by tremie, the cement content shall be not less than $380 \, kg/m^3$. Where the pile will be exposed to sea water the cement content shall not be less than $400 \, kg/m^3$. The maximum free water/cement ratio shall not exceed 0.6 for structural concrete.

The concrete composition shall be chosen to provide adequate protection against aggressive ground conditions, unless protected by permanent lining, in accordance with BS 8500-1 and BRE Special Digest 1.

The concrete composition shall be chosen to minimize bleed by consideration of free water content, fine materials content, aggregate properties and grading, and use of additions and admixtures.

Complete information on the mix composition and sources of aggregate for each concrete, the free water/cement ratio and the proposed degree of consistence, in accordance with Clause 5.2 of BS

Table B19.1 Chloride content classes and limits for concrete

	Chloride content class	Maximum chloride content (Cl⁻) by mass of cement
Unreinforced concrete	Cl 1.0	1.0%
Reinforced concrete not made with cement conforming to BS 1047 (SRPC)	Cl 0.40	0.40%
Reinforced concrete made with cement conforming to BS 1047 (SRPC)	Cl 0.20	0.20%
Pre-stressed concrete and heat-cured reinforced concrete	Cl 0.10	0.10%

8500-1, shall be provided to the Engineer before commencement of the Works.

B19.2.3 Special requirements for non-structural concrete and self-hardening slurry

Materials for use in non-structural concrete (e.g. infill piles) and plastic concrete (e.g. cut-off walls) shall meet the requirements of BS 8500 but the limits for composition of combinations shall not apply.

B19.2.4 Consistence

The Contractor shall submit the proposed consistence class or target value before commencement of the Works.

The concrete shall have sufficient consistence to enable it to be placed and compacted by the methods used in forming the piles, but without excessive bleed.

B19.2.5 Alkali–silica reaction

In order to ensure sufficient resistance to alkali–silica reaction (ASR), the concrete shall comply with the requirements of Clause 5.2 of BS 8500-2. Evidence of compliance shall be provided before commencement of the Works.

B19.3 Ready-mixed concrete

Ready-mixed concrete shall be specified, produced and transported in accordance with BS 8500. The general requirements in Clause B19.1 shall apply to ready-mixed concrete.

Ready-mixed concrete shall be produced at a plant with current certified third party accreditation to the requirements of Clause 6.4.2 of BS 8500-2. Facilities shall be provided for the Engineer to inspect the concrete batching plant or plants when requested.

All the constituents for each concrete mix shall be added at the batching plant. No extra water or other material shall be added after the concrete has left the plant except when expressly approved by the Engineer.

Each load shall be accompanied by a delivery note in accordance with Clause 11 of BS 8500-2 including the time of mixing and stating the consignee.

B19.4 Site-batched concrete
B19.4.1 General

Site-batched concrete shall be specified, produced and transported in accordance with BS 8500. Site batching plants shall be operated under a quality assurance system to ensure compliance with the requirements of BS 8500. The general requirements in Clause B19.1 shall apply to site-batched concrete.

B19.4.2 Materials
B19.4.2.1 Cement and additions

Cement and combinations shall be in accordance with the requirements of Clause 4.2 of BS 8500-2. When forwarding the piling method statement and programme to the Engineer, the Contractor shall submit details of the type of cement, or the type and composition of combination, proposed for use. High-alumina cement shall not be used.

Type II additions (fly ash, ggbs or silica fume) for use in combinations shall be in accordance with clause 4.4 of BS 8500-2.

Cement and additions shall be used in the order they are received on site.

All cement and additions shall be stored in separate containers according to type, in waterproof stores or silos.

B19.4.2.2 Aggregate

Aggregates shall consist of naturally occurring material in accordance with BS EN 12620 and PD 6682-1 unless otherwise specified or permitted. The Contractor shall inform the Engineer of the source of supply of the aggregates before the commencement of Works

Table B19.2 Limit of shell content of aggregate

Nominal maximum size of aggregate	Maximum shell content as calcium carbonate (CaCO₃) percentage by weight of dry aggregates	BS EN 12620 category
Over 10 mm	10	SC_{10}
10 mm	10	SC_{10}
4 mm	20	SC_{NR}

and, at the request of the Engineer, provide evidence regarding their properties and consistency.

Where shell is present in the aggregate, the content shall be limited as shown in Table B19.2.

All aggregates brought to the Works shall be free and kept free from deleterious materials that may affect the strength or durability of the concrete. Aggregates of different types and sizes shall be stored separately in different hoppers or different stockpiles.

B19.4.2.3 Water

Water for the Works shall comply with the requirements of BS EN 1008.

When required by the Engineer, the Contractor shall arrange for tests of the water for the Works to be carried out in accordance with BS EN 1008 before and during the progress of the Works.

B19.4.2.4 Admixtures

Admixtures complying with the requirements of BS EN 934-2 may be used.

No admixtures shall be used which contain more than the equivalent of 0.02% of anhydrous calcium chloride by weight of the cement in the concrete.

B19.4.3 Batching concrete
B19.4.3.1 General

Facilities shall be provided for the Engineer to inspect the concrete batching plant or plants when requested.

B19.4.3.2 Accuracy of weighing and measuring equipment

The weighing mechanisms and water and admixture-dispensing mechanisms shall be maintained at all times to within the limits of accuracy described in BS 1305 or such that the tolerances in Clause B19.4.3.3 are obtained and maintained.

B19.4.3.3 Tolerance in batching

Cement, additions, aggregates and water shall be batched to ±3% of the required mass; admixtures and additions added at less than 5% by mass of the cement shall be batched to ±5% of the required mass in accordance with Table 21 of BS EN 206-1.

B19.4.3.4 Moisture content of aggregates

The moisture content of aggregates shall be measured immediately before mixing and as frequently thereafter as is necessary to maintain consistency and other specified requirements. Batch weights of aggregates and water shall be adjusted to take account of the moisture content of the aggregates.

B19.4.4 Mixing concrete
B19.4.4.1 Type of mixer

The mixer shall be of the batch type, and shall have been manufactured in accordance with BS 1305 or shown by tests in accordance with BS 3963 to have mixing performance within the limits of Table 5 of BS 1305. The mixer shall be capable of achieving a uniform distribution of the constituent materials, and a uniform consistence of the concrete within the mixing time and within the mixing capacity, as required by Clause 9.6.2.3 of BS EN 206-1.

B19.4.4.2 Tolerance of mixer blades

The mixing blades of pan mixers shall be maintained within the tolerance specified by the manufacturers of the mixers, and the blades shall be replaced when it is no longer possible to maintain the tolerances by adjustment.

B19.4.4.3 Cleaning of mixers

Mixers that have been out of use for more than 30 minutes shall be thoroughly cleaned before another batch of concrete is mixed. Unless otherwise specified by the Engineer, the first batch of concrete through a mixer shall contain the normal batch quantity of cement and sand, but only two-thirds of the normal quantity of coarse aggregate. Mixing plant shall be cleaned thoroughly between the mixing of different types of cement.

B19.4.4.4 Minimum temperature

The temperature of fresh concrete shall not be allowed to fall below 5°C. No frozen material or materials containing ice shall be used except as ice-chips in mix-water to reduce the temperature of fresh concrete in hot weather, as permitted by Clause 6.2.3.2 of BS EN 1536.

B19.4.5 Delivery ticket

Each load shall be accompanied by a delivery note in accordance with Clause 11 of BS 8500-2.

B19.5 Trial mixes
B19.5.1 General

Where insufficient previous production data is available from concrete of the same composition and constituent materials, trial mixes shall be prepared for each strength class or different composition of designed concrete in accordance with Annex A of BS EN 206-1 and Clause B19.5.2. Sampling and testing of the concrete shall be in accordance with BS EN 12350 and BS EN 12390, respectively.

B19.5.2 Preliminary trial mixes

When required in accordance with Clause B19.5.1 the Contractor shall, before the commencement of concreting, have preliminary trial mixes prepared, preferably under full-scale production conditions or, if this is not possible, in a UKAS accredited laboratory using a sufficient number of samples to be representative of the aggregates and cement or combination to be used. The results of trial mixes shall be obtained from three separate batches of concrete made using the proposed composition and constituent materials.

The consistence of each trial batch shall be determined and shall be within the tolerances stated in Table 18 of BS EN 206-1 or as otherwise agreed between the Contractor and the concrete producer. The bleeding of each batch shall be measured in accordance with BS EN 480-4.

Six cubes shall be made from each batch. Two each shall be tested at 7 and 28 days and two shall be retained for further testing as necessary. In order for the mix to be accepted, the average strength of the two 28 day cubes shall exceed the specified characteristic strength by not less than $11.5 \, \text{N/mm}^2$. Alternatively, earlier tests on two cubes shall demonstrate that the specified characteristic strength at 28 days will be exceeded by at least $11.5 \, \text{N/mm}^2$.

When accelerated testing is proposed for works cubes, an additional four cubes from each batch shall be made, cured and tested in accordance with the accelerated regime.

B19.5.3 Trial mixes during the Works

Where a trial mix is required after commencement of the Works, the procedure in Clause B19.5.2 shall be adopted for full-scale production conditions. The strength requirement shall be as in Clause B19.5.2.

B19.5.4 Consistence

The consistence of each batch of the trial mixes shall be determined by the slump test as described in BS EN 12350-2 or flow table to BS EN 12350-5 and as specified in the Project Specification.

B19.5.5 Variations in composition

No variations outside the limits set out in Clause B19.4.3.3 or Table 21 of BS EN 206-1 shall be made in the proportions, the original source of the cement, additions and aggregates, or their type, size or grading class without demonstrating compliance with this Specification.

B19.6 Transporting concrete

Concrete shall be transported in uncontaminated watertight containers in such a manner that loss of material and segregation are prevented.

B19.7 Cold weather precautions

The temperature of fresh concrete shall not be allowed to fall below 5°C. In cold weather when the ambient air temperature is less than 5°C, the heads of newly cast piles are to be covered to protect them against freezing unless the final cut-off level is at least 0.25 m below the final head level as cast. Where a pile is cast in frozen ground, appropriate precautions shall be taken to protect any section of the pile in contact with the frozen soil where this occurs below the cut-off level. In the construction of pre-cast piles the requirements of Clause A.9.2 of BS 8500-1 shall be observed.

B19.8 Testing concrete
B19.8.1 Sampling

Concrete for piles shall be sampled in accordance with BS EN 12350-1 or, for consistence testing, Clause B.2.1 of BS 8500-1. Samples shall be taken from the mixer or delivery vehicle at the point of placing the concrete.

B19.8.2 Consistence testing

The consistence of concrete shall be determined by the slump test as described in BS EN 12350-2 or flow table to BS EN 12350-5 and as specified in the Project Specification and in accordance with Annex B of BS 8500-1. The consistence of each batch shall be measured and reported to the Engineer.

B19.8.3 Sampling for compressive strength testing

For each strength class or different composition of concrete, a minimum of four cubes shall be made from each sample. Unless specified otherwise in the Project Specification, a minimum number of samples shall be taken as follows:

(a) each of the first three piles on a site
(b) one sample per shift
(c) every 75 m^3 of concrete cast during the same shift
(d) at least one sample for each pile requiring concrete of strength class C35/45 or above
(e) two additional samples after interruptions of the Works longer than 7 days.

B19.8.4 Compressive strength testing

The cubes shall be made, cured and tested in accordance with the appropriate parts of BS EN 12390. Testing shall be carried out by a UKAS accredited laboratory. One cube shall be tested at an age of 7 days, two at 28 days, and one cube shall be held in reserve for further testing as required. The mean of two cube results at 28 days shall constitute a test result provided the difference of the individual cube results is not greater than 15% of the mean. The Contractor shall submit certified copies of the results of all tests to the Engineer.

B19.8.5 Acceptance criteria for compressive strength

The compressive strength shall be deemed to comply with the specified characteristic strength when both the following conditions are met:

(a) The mean strength determined from the first two, three or four consecutive test results, or from any group of four test results complies with the appropriate limits in column A of Table B19.3.

Table B19.3 Acceptance criteria for compressive strength

Group of test results	A		B
	The mean of the group of test results exceeds the specified strength class by at least: (N/mm²)		Any individual test result is not less then the characteristic strength less: (N/mm²)
First two	1		3
First three	2		3
Any consecutive four	3		3

(*b*) Any individual test result complies with the appropriate limits in column B of Table B19.3.

Characteristic strength of concrete shall mean that value of cube strength below which no more than 5% of the test results for each concrete strength class will fall as denoted by the concrete strength class in accordance with Clause B19.2.1.

19.8.6 Records of tests

The Contractor shall keep a detailed record of the results of all tests on concrete and concrete materials. Each test shall be clearly identified with the piles to which it relates.

B19.9 Steel reinforcement
B19.9.1 Condition of reinforcement

Steel reinforcement shall be stored in clean conditions. It shall be clean, and free from loose rust and loose mill scale at the time of fixing in position and subsequent concreting.

B19.9.2 Bending of reinforcement

Reinforcement shall be bent in accordance with BS 8666 to the correct shape before fixing and placing. Samples for bend tests if required will be selected by the Engineer.

B19.9.3 Fabrication of reinforcement

Reinforcement shall be in accordance with BS EN 1536, BS EN 1538 or BS EN 12794.

Reinforcement in the form of a cage shall be assembled with additional support, such as spreader forks and lacings, necessary to form a cage which can be lifted and placed without permanent distortion. Intersecting bars shall be fixed together. Hoops, links or helical reinforcement shall fit closely around the main longitudinal bars and be bound to them by wire, the ends of which shall be turned into the interior of the pile. Tying of the reinforcement shall comply with the requirements of BS 7973-2. Reinforcement shall be placed and maintained in position to provide the specified projection of reinforcement above the final cut-off level.

B19.9.4 Spacers

Spacers shall be in accordance with BS 7973-1 and shall be of durable material which will not lead to corrosion of the reinforcement or spalling of the concrete cover. Details of the means by which the Contractor plans to ensure the correct cover to and position of the reinforcement shall be submitted to the Engineer and be in accordance with BS 7973-2.

B19.9.5 Welding of reinforcement

All site welding shall be in accordance with BS EN 1011. Cold-worked reinforcement shall not be welded.

B19.9.6 Non-ferrous reinforcement

Any requirement for the use of non-ferrous reinforcement shall be included in the Project Specification.

B19.10 Grout
B19.10.1 General

(*a*) The mix design and consistence of grout to be used in the formation of piles shall produce a grout which is suitable for pumping.

Specification for piling and embedded retaining walls. Thomas Telford, London, 2007

(*b*) Grout shall consist only of Portland cement (CEM I) or blended cement to BS EN 197-1, type II additions (if required), water, fine aggregate (if permitted) and permitted admixtures. Admixtures containing chlorides shall comply with Clause B19.4.2.4.

(*c*) Fine aggregate shall be in accordance with the limits of grading 0/4(MP) or 0/4(C/P) given in Table D.1 of PD 6682-1.

(*d*) Grout shall not bleed in excess of 2% after 3 hours, or a maximum of 4%, when measured at 18°C in a covered glass cylinder approximately 100 mm in diameter with a height of grout of approximately 100 mm.

(*e*) Grout for structural purposes shall have a maximum water/cement ratio of 0.5 and a minimum cement content of $340 \, kg/m^3$. Grout for use in marine conditions shall have a minimum cement content of $400 \, kg/m^3$.

(*f*) Grout containing fine aggregate should meet the requirements of Clause 5.2 of BS 8500-2: 2006, resistance to alkali–silica reactions, unless it can be otherwise demonstrated that the risk of damaging alkali–silica reactions is acceptably low.

B19.10.2 Batching

The weighing and water-dispensing mechanisms shall be maintained at all times to within the limits of accuracy described in BS 1305 or such that the tolerances below are obtained and maintained.

The weights of fine aggregate, cement and additions for structural grouts shall be within 3% of the respective weights per batch after due allowance has been made for the presence of free water in the aggregates. The quantity of admixture shall be within 5% of the required quantity.

The moisture content of aggregates shall be measured immediately before mixing and as frequently thereafter as is necessary to maintain consistency. Necessary adjustment shall be to the weights of aggregate and water to take account of the moisture content of the aggregate.

B19.10.3 Mixing

Cement grouts shall be mixed thoroughly to ensure homogeneity.

The grout shall be mixed on site and used within 30 minutes. Following mixing the grout shall be passed through a 5 mm aperture sieve.

B19.10.4 Transporting

Grout shall be transported from the mixer to the position of the pile in such a manner that segregation does not occur.

B19.10.5 Compressive strength testing of structural grout

Cube strength testing shall be carried out in accordance with BS EN 12390.

Where grout is supplied ready mixed, sampling requirements are as in Clause B19.8.3. For site-batched grout, one sample shall be taken from each $15 \, m^3$ of grout or part thereof in each shift and shall consist of a set of six 100 mm cubes. Two cubes shall be tested at 7 days and two at 28 days after casting. The remaining two cubes shall be retained for test at a later age, as and when required or as specified in the Project Specification. The mean of two cube results at any test age shall constitute a test result provided the difference of the individual cube results is not greater than 15% of the mean. The Contractor shall submit certified copies of the results of all tests to the Engineer.

B19.10.6 Acceptance criteria for compressive strength of grout

The acceptance criteria for compressive strength of grout shall be the same as those for concrete given in Clause B19.8.5.

Table B19.4 Form for specification of designed concrete

1	Concrete designation				
2	Strength class				
3	DC-class				
4	Maximum water/cement ratio				
5	Minimum cement content (kg/m^3)				
6	Permitted cement and combination types				
7	Nominal maximum size of aggregate (mm)				
8	Chloride class				
9	Special requirements for cement or combination				
10	Special requirements for aggregates				
11	Special requirements for temperature of fresh concrete				
12	Special requirements for strength development				
13	Special requirements for heat development during hydration				
14	Other special technical requirements				
15	Additional requirements				
16	Rate of sampling for strength testing				
17	Use of recycled aggregate (RCA) permitted	Yes/No	Yes/No	Yes/No	Yes/No
18	Target consistence or consistence class				
19	Tolerance on target consistence if different from BS EN 206-1				
20	Method of placing concrete				

B19.10.7 Placing grout in cold weather

Grout shall have a minimum temperature of 5°C when placed. No frozen material or material containing ice shall be used for making grout. All plant and equipment used in the transporting and placing of grout shall be free of ice that could enter the grout.

B19.11 Form for specification of designed concrete

Designed concrete and structural grout shall be supplied in accordance with BS 8500-2, and Table B19.4.

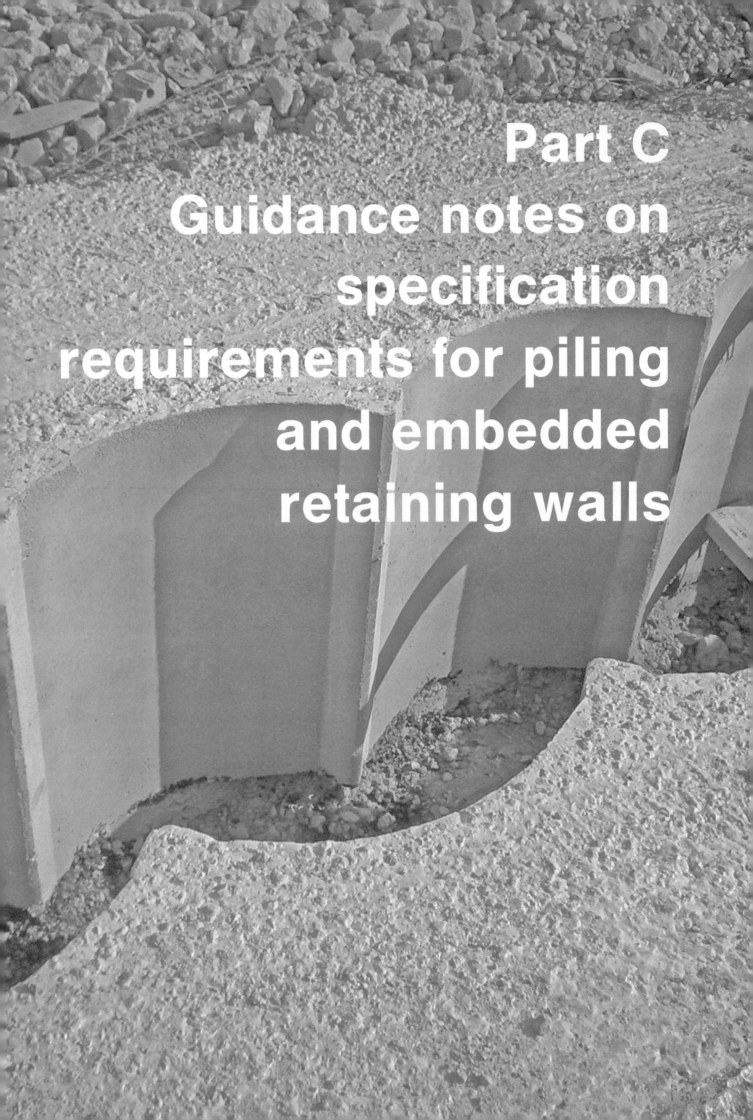

**Part C
Guidance notes on specification requirements for piling and embedded retaining walls**

Guidance notes on specification requirements

C1 Specification Requirements for piling and embedded retaining walls

C1.1 Standards

This revision of the Specification has been published at a time when UK industry is in the transition to using European Standards. Some European Standards have already been published and some are still being drafted. British Standards and European Standards are expected to co-exist until about 2010 although the timescales are not yet clear. Specifiers will need to monitor the status of all standards specified.

Due consideration should be given to precedence in UK practice, and the Contractor can be asked to demonstrate precedence using previous projects of a similar foundation type in similar ground conditions.

C1.2 Project Specification

The document has been assembled so that the specifier selects the clauses of the Specification considered to be appropriate for the Works and completes the appropriate Project Specification for each clause selected. Section B1 will be required in all cases. Section B19 will be required for all piles or embedded retaining walls made of concrete. The other sections will be selected depending on the particular requirements of the structure, site and environmental constraints. In many cases there will be more than one type of pile or wall that could provide an appropriate solution and the Engineer may choose to allow the Contractor to select from a range of possible types. A list of the requirements to be specified is given in the Project Specification.

Minipiles are not implicitly covered in this document as minipiles can be constructed using a plethora of techniques. However, it should be possible to specify minipiling techniques using this document, with a Project Specification combined with all, or part of, one or more of Sections B2 to B7.

Other requirements can also be specified by adding sections to the Project Specification. For example, as sustainable construction is an important issue for foundation works if sites are not to become permanently sterilized, requirements for the removal of piles at the end of their life could be specified.

(a) Role of the Engineer
If it is not the Engineer, the body (Architect, Contract Administrator, etc.) acting as supervising officer should be specified in the Project Specification. If the powers of that body are not clearly defined elsewhere in the contract, then details should be provided here.

(h) Submission of information
Refer to Clause C1.12.

(q) Performance criteria for piles under test or wall elements during service
The settlement of a single pile under load depends on the proportion of load carried by the shaft and base, the stiffness of the soil below the base, and on the elastic shortening of the pile section. The settlement of working piles may also be influenced by pile group effects. The requirements of the structure dictate the tolerable settlement and differential

settlements. These factors depend in turn on the structural form, materials, finishes, load distribution and variations, and aesthetics.

It is therefore not surprising that no simple rule for expected settlement can be established without taking the various factors into account. Setting criteria in a wholly logical and rational way can, therefore, be time consuming and could demand a level of analysis that is unwarranted. As a result it is common to find that limiting criteria have been chosen on the basis of some earlier specification, company practice or structural advice based on the concept that foundations should behave as rigid supports.

For most ordinary steel or reinforced concrete framed buildings, individual pile settlements of perhaps 5 mm to 10 mm at working load would be normal and acceptable, with perhaps 15 mm to 25 mm at $1.5 \times$ working load. Greater settlements can be expected for long thin piles, where elastic shortening may become significant, or for very heavily loaded piles. Specific situations, such as where piles with very different loading conditions are closely spaced or where load reversals are possible, should always be borne in mind and appropriate values selected, albeit they may imply deeper or larger piles and additional cost.

C1.4 Design

There has been confusion in the past over the division of design and construction responsibilities, leading to disputes. Clarity has been lost where:

- responsibilities have been left to be inferred from the Specification and/or the contract/subcontract, rather than being clearly stated
- conflicting responsibilities have been stated in both the Specification and the contact/subcontract documents
- the Contractor chooses to involve his own engineer and/or a specialist contractor to undertake specific design duties without clarifying the specific roles and responsibilities of each party.

To reduce the number of disputes in the industry, this Specification is not intended to allocate or imply any responsibilities for design and the risks associated with design. These must all be comprehensively addressed in the contract/subcontract documents. Advice on how this can be achieved is available at www.fps.org.uk.

This Specification is intended to be used with any form of contract, thus it is imperative that the division of design responsibilities are clearly stated in the enquiry and contract documents. The specifier should decide who is to be responsible for the design of the foundation scheme, the choice of the piling or embedded walling method, and the design of the piles or wall elements. To help clarify the division of the scope and extent of these responsibilities it is strongly recommended that a table, such as Table C1.1, is provided in the enquiry documents and incorporated into the contract documents.

Basic options are provided in Clause B1.4 for either the Engineer or the Contractor to design the piles or the wall elements. The Engineer can either provide the design (Option 1) or specify that the Contractor designs the piles (Option 2) or the wall elements (Option 3). This change was introduced to clarify design responsibilities. Further clarification will be necessary when there are many parties to the design. These design options do not correspond with those published in earlier versions of this Specification.

Where the Engineer has designed the piles or wall elements, the assumptions and criteria used in the design should be stated in the

Table C1.1 Design responsibilities

Design responsibility	Engineer	Contractor
1. Design of foundation scheme (including SWL and pile location)		
2. Choice of piling or walling method		
3. Design of piles or wall elements to carry Specified Loadings		

enquiry and contract documents. If the Contractor considers the Engineer's design to be inappropriate, this should be stated in the tender.

If the Engineer decides to delegate responsibility for the design of the foundation scheme, choice of the piling or walling method, and/ or design of the piles or wall elements to the Contractor, then a Performance Specification should be provided. The design and performance criteria, as appropriate, should be defined in the Performance Specification, which may include:

- Design standards to be adopted for both geotechnical and structural design.
- Settlement requirements for individual piles when the Contractor is required to design the piles.
- Settlement requirements for pile groups (only when the Contractor is required to design the foundation scheme).
- Permissible deflections for embedded walls when the Contractor is required to design the wall elements.
- Limits on building settlement and distortion (only when the Contractor is required to design the foundation scheme).
- Minimum testing requirements.
- Working Loads or Representative Actions (compression, tension, horizontal, bending moments).
- Ultimate Loads or actions where appropriate.

Where the Engineer chooses not to be responsible for the design of the support of the individual loadings, i.e. Contractor design of piles, the Engineer will need to consider soil–structure interaction and pile group effects in arriving at the performance requirements for an individual pile under test. It is advisable that the Engineer maintains flexibility in the design of the foundation scheme during the tender period. The design of the substructure should be able to accommodate a reasonable range of choices by the Contractor. The Contractor should be allowed, where possible, to offer any solution that meets the specified performance requirements. The Contractor should include in their tender the testing regime that will demonstrate that the specified performance has been provided.

When responsible for determining the method for constructing the piles, the Contractor may then also become responsible for the design of the piles whilst the Engineer retains full responsibility for the design of the foundation scheme. This process of delegation should facilitate the use of new, innovative or alternative piling methods.

Allowing the Contractor to select the walling method encourages comparisons of the benefits of concrete solutions, such as pile or diaphragm walls, against those of steel solutions, such as sheetpiling or combi-walls.

Unless stated otherwise in the contract, it is likely that whoever is responsible for selecting the piling or walling method will become responsible for any associated risks which are reasonably foreseeable, which may include:

- Installation induced ground movements (settlement and heave) and their impacts on structures.
- Noise, dust and vibrations.
- Impacts on pile trimming.

Sharing particular responsibilities with the Contractor will not relieve the Engineer from his overall scheme responsibilities.

Design (guidance for piles)

Option 1 — Engineer design of piles

Where the Engineer prepares a design (Option 1), the piling type selected should be based upon the use of a non-proprietary system. The following information, as a minimum, should be shown on the drawings (and cross-referenced in the Project Specification):

- Pile layout.
- Working Loads (or Representative Actions).
- Ultimate Loads where applicable.
- Location of any preliminary or working test piles.

In these instances practical advice should be sought from Contractors, or from suppliers, prior to finalizing the design, since the pile design parameters are often substantially influenced by the method of construction. Normally such advice would be non-contractual and should be evaluated accordingly. The basis of the design should be made clear in the enquiry documents so that each Contractor can select a suitable approach.

Decisions made early in the design process frequently have a profound influence on the successful construction of piled foundations (e.g. the selection of pile method and diameter). It is important that the Engineer recognizes that these decisions reflect not only the structural requirements of the works but also constraints arising from the environment (e.g. site size, location and topography, access restrictions) and from buildability (e.g. member size, reinforcement congestion).

It is vital that the method of construction is compatible with the Engineer's intentions. Early discussions between the Engineer and Contractor are helpful in achieving this and may avoid inappropriate techniques at the specification stage. Nevertheless, prior to construction the Engineer must be satisfied that the proposed method of construction will achieve the requirements. This assessment cannot reliably be made on structural grounds alone. Input from a suitably experienced geotechnical engineer is essential. It may also be necessary to carry out load testing to prove that the method of construction is appropriate.

Disclosure of the design philosophy at tender stage, and ongoing open discussion throughout tender evaluation and construction will help to ensure the quality of the end product. At tender stage, without such information, the Contractor will have only a restricted basis upon which to judge the suitability of the construction method. During construction the Contractor will be less able to identify and highlight factors which may influence the parameters of the design.

During construction, unexpected and emergency situations may arise which, if not dealt with rapidly, could jeopardize the integrity or performance of the completed foundation. It is essential for foundation construction that the contractual arrangements between the various parties accept the possibility that unexpected conditions can occur and that remedial action can be agreed sufficiently quickly for the work to continue without any adverse effects on the completed foundation. For this reason full-time supervision by the Engineer and the Contractor is recommended for all piling works. These site representatives should be sufficiently familiar with pile construction and the design requirements to enable solutions to unexpected or emergency situations to be agreed without unreasonable delay.

Option 2 — Contractor design of piles
If Option 2 is selected then the Engineer must provide performance criteria for the piles and sufficient data on the loads and moments (actions) to be applied to the individual piles for the Contractor to be able to assess the total requirements for the piles.

There are generally two categories of Contractor designed piles:

1. Pile layout as designated by the Engineer.
2. Alternative pile layout to suit the Contractor's design.

For category 1, the Engineer provides the information required for setting out the piles, but for category 2 the Contractor should provide this information.

The Engineer should provide the overall performance requirements for the structure, including the settlement requirements of pile groups, if the Contractor is responsible for the design of the foundation scheme.

In many cases the Engineer will design the foundation scheme, including pile positions and the substructure (including the pile caps, ground beams etc.) and the Contractor will only design the individual piles to achieve the criteria specified by the Engineer. The Engineer will nevertheless be expected to take overall responsibility for the design of the foundation and in some cases may be required to adopt any Contractor design elements. Where the design responsibilities are divided, clear communication of all the design requirements is essential. Consideration should be given to defining the design criteria within the Project Specification, which may include the following:

- Design standards to be adopted for geotechnical and structural design.
- Available geotechnical data.
- Working Loads (or Representative Actions).
- Ultimate Loads where applicable.
- Minimum factor of safety.
- Settlement requirements for individual piles.
- Minimum pile testing requirements.
- Structural performance and settlement requirements for pile groups, where applicable.

The Contractor will aim to provide the most economic solution within the specified design constraints. It is therefore important to ensure that the constraints are realistic and that they are the minimum requirements necessary to achieve the overall foundation performance. Unrealistic settlement or lateral movement criteria, unnecessary restrictions on pile diameter, or excessive requirements for cleaning pile bases, for example, can substantially increase pile costs.

The Engineer should obtain from the Contractor full details of the type of pile offered, the standards of control to be used, how the calculation and checking of the load-carrying capacity and settlement of the piles will be carried out, and the proof loading or other testing proposed to be undertaken on the site.

The following notes should be followed when pile performance is being specified in Table B1.2. Each note corresponds to a column in the table, reading from left to right:

- Each and every pile should be allocated a unique reference number or code.
- Permitted types of pile are restricted to those specified in the corresponding section of the Specification, e.g. insertion of 'Section B3' will restrict permitted type(s) of pile to bored cast-in-place piles only. Further specification, e.g. 'underreams not permitted' may be necessary in particular circumstances.
- Working Loads specified on drawings should be grouped and each group allocated to one Allowable Pile Capacity. This reduces the number of different pile sizes on site and helps eliminate the confusion that may arise when each pile is individually sized.
- The pile designation relates to the size of the particular pile and possibly to the reinforcement cage. The number of different pile sizes for the project will then be listed separately.
- The DVL may be much larger than the Allowable Pile Capacity.
- The factor of safety need not be specified as this is a Performance Specification. However, it is acceptable for the Engineer to specify a minimum factor of safety if deemed appropriate.
- A realistic estimate of the likely lower bound load settlement curve for a pile tested in isolation should be made, and the permitted settlement at 100% DVL taken from that curve. If a stiffer pile than is indicated by the permitted settlement is required, then the pile type or dimensions will have to be changed (e.g. pile lengthened or a larger size).
- The stratigraphy of the ground may make it imperative that piles have a minimum length, to ensure penetration into a particular stratum; alternatively the minimum penetration into a particular stratum can be specified (the column heading will then require to be changed to 'Minimum penetration into XYZ').
- Minimum pile dimensions will normally be determined by the permitted stresses in the pile materials, taking account of axial loads, moments and transverse loads. Maximum pile dimensions may be determined from pile spacings.

Table B1.2 has remained substantially unchanged from the previous (1996) version of the Specification. With the future introduction of European Standards and the withdrawal of current national standards, it is expected that the table headings will not tally with the terminology used in the new codes. In order to define pile loads for use with BS EN 1997 it will be necessary to state the Representative Actions for each pile, and the definition of Specified Working Load will not be relevant. It is expected that use of this table will evolve, as the UK becomes familiar with the new codes, and as such it is not desirable to change the current format at this time. If a BS EN 1997 design approach is to be used, it is envisaged that the Engineer will devise their own version of this table for use in the Project Specification.

Design (guidance for walls)

Options are provided in Clause B1.4 for either the Engineer or the Contractor to design an embedded retaining wall. Reference should also be made to BS EN 1997-1 and CIRIA C580. Careful consideration needs to be given to the allocation of design responsibilities for walls. Any basement construction normally comprises four main

Table C1.2 Main basement construction activities

Activity	1. Wall construction	2. Excavation process	3. Construction of substructure	4. Construction of superstructure
Input data	1. Site Investigation 2. Architects plan 3. Depth of basement 4. Assumed excavation and propping sequence	1. Shape of formation level 2. Detailed levels of basement floors 3. Finalized excavation and propping sequence	1. Final layout of superstructure 2. Final loads 3. Final levels	1. Full design
Construction operations	1. Install wall	1. Trimming of wall 2. Construct capping beam 3. Dewatering 4. Excavation 5. Propping 6. Prepare formation	1. Piling for superstructure 2. Construct basement slab 3. Construct basement floors	1. Construct superstructure
Potential effects and risks	1. Ground movements due to installation 2. Lack of watertightness	1. Lowering of groundwater outside of basement 2. Movement of surrounding ground 3. Water inflow either through wall or beneath wall	1. Additional ground movements when props removed 2. Internal water control measures	1. Superstructure Contractor's acceptance of basement

activities; wall construction, excavation process, construction of the substructure and construction of the superstructure. Each of these activities requires different forms of expertise, much of which will lie outside a walling contractor's capability. The Engineer should consider the influence of ground movements caused by work activities which precede piling operations. Activities which may influence the works include demolition of existing structures or excavation of on-site material prior to walling/piling. The Engineer should provide details of such activities in the Project Specification to a Contractor tendering for the design of elements, as they may affect the design of for example; heave steel to prevent the formation of cracks in bearing piles. The influence of ground movements caused by such activities should also be considered in the setting of appropriate limits on movements for both new elements to be constructed as part of the works, and neighbouring structures. Each activity requires different input data, has different construction operations and has different effects and risks. These activities are also interdependent and have interfaces that need to be carefully managed. The key issues are summarized in Tables C1.2 and C1.3.

The design of the wall should take into account the following:

- Required design life.
- Durability: crack control in reinforced concrete depending on the exposure conditions, the strength and permeability of the mix for

Table C1.3 Advantages and disadvantages of main basement construction activities

Combinations of activities	Advantages	Disadvantages
Activity 1 only	1. Can start work on wall construction before detailed design	1. Many assumptions dependent on undeveloped aspects of design 2. Divided key responsibilities at interfaces between activities 1 and 2
Activities 1 and 2	1. Single point of responsibility for basement construction	1. Much of information for activity 2 may not be available
Activities 1, 2 and 3	1. Easier to co-ordinate all activities	1. Needs full design complete before starting 2. Responsibility is divided with activity 4

infill secant piles, the corrosion of steel piles, the deterioration of timber lagging used for king post walls, etc.

- Sequence of construction: the sequence of temporary or permanent propping, extent of stages of excavation and timings.
- Drainage: whether the penetration of the wall is to restrict groundwater flow and any provisions for drainage in front of the wall at formation level.
- Watertightness: the Designer and Employer need to have a clear understanding of what is required for the permanent works and what is reasonably achievable with the different forms of wall construction.
- Structural loads and moments (actions) applied to the wall.
- Soil and water pressures: these will vary according to the sequence of construction, temporary and permanent wall restraint and the effects of dewatering, over-excavation, erosion, etc.
- Effects on surrounding structures: adjacent buildings, roads and utilities may be affected by ground movements associated with wall construction and the process of excavation in front of the wall. Vibration can cause settlements of loose granular soils, and dewatering may cause consolidation of cohesive soils and draw fines from cohesionless soils. Deflection of the wall during excavation of soil in front of the wall will cause settlement and horizontal movements of the retained soils.

There are many design choices that will have an inherent impact on the design of the permanent works. For example, the choice of a ground bearing basement slab may result in the need for long term heave forces to be withstood by the structure, and the choice of underslab drainage may reduce the passive resistance available for the wall. It is recommended that basement design is only undertaken by a suitably experienced engineer.

Option 1 — Engineer design of walls
Where the Engineer prepares a design (Option 1) the wall type selected should be based upon the use of a non-proprietary system. The following information should be shown on the drawings (and cross-referenced in the Project Specification):

- Wall layout.
- Depth and thickness of the wall.
- Propping forces provided by elements of the structure.
- Design life of the wall.
- Bending moment and shear force envelopes (if the Contractor is detailing the reinforcement).
- Typical reinforcement details or steel pile section moduli as appropriate.
- Soil design parameters and design groundwater levels.
- Required box-out and shear connection details.
- Required watertightness.
- Limits on wall and/or ground movements, vibration and noise.

In these instances practical advice should be sought from Contractors prior to finalizing the design, since the wall design parameters can be substantially influenced by the method of construction. Normally such advice would be non-contractual and should be evaluated accordingly. The basis of the design should be made clear in the enquiry documents so that each Contractor can select a suitable approach.

Decisions made early in the design process frequently have a profound influence on the successful construction of walls (e.g. the selection of walling method and sizes). It is important that the Engineer recognizes that these decisions reflect not only the structural requirements of the works but also constraints arising from the environment (e.g. site size, location and topography, access restrictions) and from buildability (e.g. member size, reinforcement congestion).

It is vital that the method of construction is compatible with the Engineer's intentions. Early discussions between the Engineer and Contractor are helpful in achieving this and may avoid inappropriate techniques and tolerances at the specification stage. Nevertheless, prior to construction the Engineer must be satisfied that the proposed method of construction will achieve the requirements. This assessment cannot reliably be made on structural grounds alone. Input from a suitably experienced geotechnical engineer is essential.

Disclosure of the design philosophy at tender stage, and ongoing open discussion throughout construction will help to ensure the quality of the end product. At tender stage, without such information, the Contractor will have only a restricted basis upon which to judge the suitability of the construction method. During construction the Contractor will be less able to identify and highlight factors which may influence the parameters of the design.

During construction, unexpected and emergency situations may arise which, if not dealt with rapidly, could jeopardize the integrity or performance of the completed foundation. It is essential for foundation construction that the contractual arrangements between the various parties accept the possibility that unexpected conditions can occur and that remedial action can be agreed sufficiently quickly for the work to continue without any adverse effects on the completed foundation. For this reason full-time supervision by the Engineer and the Contractor is recommended for all walling works. These site representatives should be sufficiently familiar with wall construction and the design requirements to enable solutions to unexpected or emergency situations to be agreed without unreasonable delay.

Engineer design for the permanent condition only
Where the Engineer opts to design the wall only for the permanent condition, the Engineer should then ensure that the Contractor undertakes a design as Option 3, but only for the temporary condition. The Contractor should submit details of the wall elements in accordance with the Project Specification along with details of any temporary works associated with the installation of the wall and subsequent excavation.

Option 3 — Contractor design of wall elements
Where the Contractor is required to design the wall elements, the Engineer should provide the following information on constraints to the design:

- Wall location.
- Permissible lateral wall deflection.
- Required watertightness of the wall above the formation level.
- Required groundwater control measures below formation level, both through the wall and through the ground.
- Design standards to be adopted for geotechnical and structural design.
- Design life of the wall.

- Any constraints on the allowable construction sequence.
- The permanent works floor slab levels, and details of lift pits, sumps and slab voids as appropriate.
- The permanent works which will incorporate the completed wall including any connections, facing details, and the loads and moments that will be applied to the wall in the permanent condition from other parts of the structure.
- Adjacent structures and utilities to be considered in the design including any limits on damage, noise and vibration, and requirements for monitoring and reporting.

C1.4.1 Wall Manual

The concept of the Wall Manual is to encourage effective communication of the key safety and technical issues between the Engineer, Contractor and Designer, especially for the interfaces between activities 1, 2 and 3 described in Table C1.2. The Contractor is encouraged to use the Wall Manual to communicate these key safety and technical issues to the site personnel undertaking the basement construction works.

It is not the intention that the Wall Manual is a cumbersome document, but is a clear and concise tool that will allow:

- The Contractor to understand the Designer's assumed excavation and propping sequence.
- The Designer to state all assumptions that need to be verified on site and who is responsible for this verification.
- The Designer to be notified by the Contractor of any changes.
- For key safety and technical risks to be communicated by the Designer to the Contractor.
- For key safety and technical risks to be communicated by the Contractor to his site teams and/or subcontractors.

The Wall Manual may be incorporated within the Geotechnical Design Report provided that the intentions above can still be achieved.

C1.6 Safety
C1.6.1 Standards

The design and construction of the works must be carried out in accordance with the Construction (Design & Management) Regulations (CDM). Some specific work sector guidance for designers can be obtained from CIRIA Report 166 (1997).

C1.6.2 Working Platform

Guidance on the design and management of working platforms for piling and embedded walling works is given in *Working Platforms for Tracked Plant* (2004) BRE BR470.

Particular guidance for hydraulically bound platforms is given by the Waste and Resources Action Programme (WRAP: see www.wrap.org.uk).

C1.7 Ground conditions

The factual information reported in the Site Investigation forms the basis of key decisions in regard to both the design and the construction of the piles or walls. The information must therefore be appropriate to the works to be carried out, full, accurate and clearly presented, so that both the Designer and the Contractor can place reliance on it. The responsibility for the accuracy of the data inevitably lies with those who exercise control over the SI (the Employer, the Engineer and the SI contractor).

A comprehensive, accurate and clearly presented SI carried out in accordance with BS EN 1997-2 is an essential prerequisite to the selection, design and construction of piles or embedded walls. Particular attention should be given to the following list of requirements of such an SI:

- A comprehensive desk study of the site history to assess the risk of obstructions, contamination, quarries, opencast and deep mines, backfilled sites or archaeological finds that could affect the feasibility of pile or wall construction, programme and cost; suspected obstructions should be identified by probing or preliminary enabling works to confirm their extent.
- Equilibrium piezometric levels of all possible water tables, including artesian conditions and any seasonal or tidal fluctuations.
- Permeability of the soils.
- Presence of coarse, open soils, cavities, natural or artificial, which may cause sudden losses of support fluid in open excavation and require preliminary treatment.
- Strength and deformation characteristics of soils, particularly weak strata that could cause instability or large deformation.
- The presence of boulders or obstructions which may cause difficulties in excavations or driving conditions.
- Soil and groundwater chemistry which may affect the durability of the piles or wall, the disposal of spoil, and the performance or disposal of support fluid.
- The strength and profile of any rock surface beneath the area to be piled, or the wall alignment, in particular a steeply inclined rockhead surface.

The SI should address both the design and construction phases of the project. The most common inadequacies include:

- Inadequate depth of investigation.
- Inappropriate drilling techniques.
- Inadequate groundwater data.
- Inadequate location data (position and level relative to the works).

It is in the interest of the project as a whole that all parties seek to communicate all relevant data, whenever possible. Having carried out a detailed SI, the information gained is used to maximum advantage by ensuring that all relevant parties have ready access to it. The Designer must be provided with a full copy of the SI reports if he is to produce a fully detailed design. The Contractor should be provided, as a minimum at tender stage, with a full copy of the GI. The flow of such information is the responsibility of the Employer, Engineer and Contractor alike.

Conditions of Contract may affect what ground information should be provided to the Contractor in SI reports and the responsibility attracted by the various parties particularly where interpretation is included in the reports. Reports given to the Contractor should include both factual and interpretative data, as part of the Geotechnical Design Report. These reports should be checked to ensure that they are relevant to the structure to be built and that the scope of construction works envisaged at the time the SI was carried out has not significantly changed. If they have, or if the SI reveals additional problems, further investigation should be carried out.

C1.8 Installation tolerances

The tolerances specified are realistic for most sites and ground conditions. However, where the ground contains obstructions, or for certain ground conditions (e.g. pile tip moving from softer to harder layers), tolerances may need to be relaxed. This will necessitate an allowance in the design of pile caps and ground beams to suit the installed pile positions. The Contractor should be informed of the reasons leading to the changed tolerances. The Engineer should ensure that the design of the structure to be built on the piles is not compromised. In some cases the design may require tighter tolerances than those specified in which case the required tolerances should be specified in the Project Specification. Plunged stanchions may need to be positioned to structural tolerances which are much tighter than those for piling. If this is the case, the tolerances should be specified in the Project Specification and the Contractor's method statement should show how the tighter tolerances are to be attained.

In most cases the tolerances quoted in the Specification are achievable. They may not, however, be practical under the specific conditions of a given project. Piling under restricted headroom, for example, may require the use of shorter tools and thereby reduce control of verticality. Similarly, piling in ground containing obstructions may make the tolerances impracticable or, at best, costly to achieve.

BS EN 1536 states different tolerances to those in this document. This Specification allows for project specific tolerances to be agreed prior to commencement of the works, to account for design requirements, ground conditions and available piling equipment. The guidance in this document should help the Designer to specify realistically achievable tolerances and thereby avoid conflict which can arise from a mismatch between the design assumptions and construction practice.

The Designer should specify only those tolerances necessary to achieve the desired quality. Unnecessarily restrictive tolerances inevitably lead to increased costs. The Designer can, in many cases, avoid the need to specify difficult or impracticable tolerances by addressing them at the design stage. This can be achieved, for example, by incorporating appropriate tolerances in the design of interfacing works such as pile caps and ground beams.

Minor deviations from specified tolerances in individual piles are rarely critical to the performance of the foundation as a whole. Where piles are constructed in large groups, for example, occasional piles constructed marginally out of tolerance are commonly counteracted by adjacent piles. Similarly, where a pile deviates from the specified tolerance at depth, it is rarely found to have any significant impact on pile performance because of the restraining influence of the ground provided the ground has a strength greater than 10 kPa.

The permitted deviation in position of a pile with a cut-off at, or above, commencing level is given as 75 mm in any direction in Table B1.4. Where cut-off level is at some depth below commencing level, the Designer should take into account the combined effects of this positional tolerance and the verticality tolerance. Using these tolerances, for example, a bearing pile with cut-off 3 m below commencing level could have an allowable deviation at cut-off level of 115 mm.

C1.8.1 Setting out

In order to minimize the risk of disturbance from the movement of the piling machine and other site traffic it is preferable to mark the pile positions using visible steel pins of sufficient length (minimum 300 mm) and diameter (minimum 12 mm).

The pin should be driven to ground level, or just below, and the location marked with a contrasting material such as a small quantity of sand. Just before the pin is driven to its final level, it should be checked for accuracy and the top painted with an agreed colour to avoid any confusion with 'line' pegs. Where the density of piling is high, it is good practice to identify the pin using a tag giving the pile reference number.

If raking piles are to be installed then the setting out pin is located in an offset position at piling platform level taking account of the depth to cut-off level and the rake of the pile. In addition an alignment pin is required to indicate the direction of rake.

The installation and checking of marker pins is an essential means of setting out control. It does not, however, preclude inaccuracies arising during the early stages of pile installation, when the original marker pin has been removed. In order to meet the specified tolerances, offset pins should be installed prior to removal of the marker pin to provide a continuing reference system, ensuring maximum accuracy in positioning of the bore and casing.

C1.8.2 *Forcible corrections to piles*

Forced corrections to piles will lead to the introduction of stresses and strains which may not have been taken into account in the design. Moving the pile at ground level will have little or no effect on the pile buried in the ground. Even when it appears that a correction has been achieved, once the correction force is removed it is likely that the pile will return to a position at, or close to, its original position. In the case of steel-sheet piles this is one of the principal causes of declutching.

C1.9 *Waterproofing of retaining walls*

This new clause represents a clarification of responsibilities for the watertightness of retaining walls. For walls built to the standards required by this Specification, the Contractor may only be responsible for the watertightness of the exposed faces of the wall, and not necessarily for the watertightness of the wall below final formation level. Responsibilities must be defined in the Project Specification, as it may be that groundwater flow beneath formation level could only be controlled by, for example, a separate dewatering system.

Leakage through walls can often occur at the boundary between the wall elements and the capping beam or slabs. The design of the waterproofing system is often inadequate at these boundaries, for example, the slab connection waterproofing membrane often does not cover the area of concrete between the slab and the water stop in a diaphragm wall panel leaving a potential water flow path. It is essential that responsibility for these interfaces is defined in the Project Specification. Detailed guidance on these interface waterproofing issues can be found in Puller (1994).

The watertightness of concrete walls can be affected by cracks and movements at the joints between piles or panels that form due to wall deflections and/or differences in support conditions, and by any openings in the wall or in the floor slabs. Even thick concrete walls will allow some transmission of water through, and a simple calculation using Darcy's Law can be used to estimate flow. The following measures will help improve the watertightness of any embedded retaining wall:

- High quality design and construction.
- Sealing all joints and repairing faulty construction (most effective after the majority of wall movements have taken place).
- The use of inner walls cast onto the embedded retaining wall together with cavity wall construction with provisions for drainage, sumps and ventilation.
- The application of interior waterproofing using membranes after the embedded walls have been repaired and sealed.
- The use of water stops cast into all joints and connections exposed to water.

Table C1.4 Waterproofing performance criteria for walls

BS8102 requirements for whole system		Corresponding criterion for retaining wall component
Grade of basement	Performance level	
1 (basic utility)	Some seepage and damp patches tolerable	Beading and limited damp patches tolerable. No weeping
2 (better utility)	No water penetration but moisture vapour tolerable	Beading and limited damp patches tolerable. No weeping. Other components will be needed in addition to the retaining wall component to achieve the required watertightness of whole system
3 (habitable)	Dry environment	
4 (special)	Totally dry environment	

Basement design and the expected watertightness of any embedded retaining wall should be carefully planned (with a suitable period allowed in the contract programme) and presented in the contract documents so as to avoid disputes and litigation.

Table C1.4 is a guide to the achievable grade of watertightness for walls built to this Specification, and their relationship with the overall system grades given in BS8102. CIRIA guide 139 also provides useful guidance on achieving the required grade of basement space.

C1.9.2 Repair

Prior to construction, it is good practice for the Contractor to determine the most suitable method(s) and timing(s) of repairs and include these in the method statement. The method of repair will include identifying the source and cause of any leak, providing access to the source of the leak, preparation work, undertaking the repair works and monitoring the effectiveness of the repair as well as its impact on the watertightness of the remainder of the wall. It is necessary to allow for these potential activities in the programme.

It is important for any repair system using injection that the injected material will be effective for its intended design life. Chemically aggressive groundwater may degrade the injected material, hence for all walling works the groundwater chemistry should be determined during the SI.

The cleaning and repair of walls should be carried out such that the integrity and performance of the wall is not adversely affected.

C1.10 Construction method

The construction method statement should include sufficient detail to demonstrate that it is compatible with the design assumptions and any site constraints. The amount of detail required at the time of tender should be considered, bearing in mind what is required for an initial assessment and comparison of tender proposals and what is reasonable for a Contractor to provide in a competitive situation. The Contractor's method statements and designs must be regarded as confidential (copyright laws apply) and parties receiving such information should, if the Contractor so requests, be prepared to sign confidentiality agreements. The following list includes examples of items that may need to be provided by the Contractor:

- site staff and organization
- the type of piles or wall elements
- the experience of the Contractor and his staff with the piling or walling type and the particular ground conditions
- the type and number of rigs and ancillary equipment
- the details of plant should be sufficient to demonstrate its suitability for achieving the required penetration and to work within any noise and vibration limits where these are restricted

Specification for piling and embedded retaining walls. Thomas Telford, London, 2007

- the sequence of piling or walling which may affect the integrity of nearby piles
- setting out and means of achieving specified tolerances
- the time period for construction for a complete pile
- the pile construction process in detail
- the methods that will be used to check pile or wall element depth and compliance with the specified tolerances
- the frequency and means of testing materials
- details of pre-boring, or other means of aiding pile driveability, and measures that will be taken to minimize disturbance of the surrounding ground
- the level to which concrete will be cast or steel installed
- material properties to meet the required design life and durability
- equipment and method for monitoring surveys including frequency of readings and the form of presentation of results if not previously specified by the Engineer
- trials in advance of the main installation to demonstrate that the method and materials will meet the Engineer's specified requirements
- the method of achieving the specified verticality and positional tolerance including details of guide walls or guide frames with possibly radar or sonar being used to prove the excavated element profile
- procedures for dealing with emergency situations such as sudden loss of support fluid, obstructions and piles that are out of tolerance
- company safety plan
- quality assurance procedures
- typical record sheets.

The above recognizes that, at tender stage, it is generally not appropriate to require detailed method statements for the complete works. However, sufficient detail should be provided by the Contractor to enable the Engineer to assess the suitability of the construction method, in relation to the design, specification and site conditions. Without placing unduly onerous requirements on the Contractor the tender documents should define the minimum information required to make this assessment. Where the Engineer is the Designer (Option 1) disclosure of the design philosophy at tender stage will also help the Contractor to select an appropriate method of construction.

C1.10.1 Obstructions and voids

Shallow obstructions in the ground are normally most economically removed by an excavator before the commencement of the piling or embedded walling works. Deeper obstructions, say deeper than 4 m, or when in areas of archaeological interest, or when close to sensitive structures, may then require the use of coring methods to penetrate or remove obstructions. There are many methods of coring/removing obstructions, all of which are likely to have noise, vibration and dust implications.

Voids in the ground can either be backfilled prior to the commencement of the works, or piles may be constructed using permanent sleeves.

C1.10.3 Concrete casting level tolerances

In order to ensure sound concrete at the pile head, it is necessary to provide a working platform at least 300 mm above the specified cut-off level. The working platform may need to be significantly higher than 300 mm above cut-off level so as to be above the reinforcement projection or to accommodate the use of guide walls for embedded walling works. In addition, the working platform level needs to be high enough above the cut-off level to accommodate the casting level tolerances.

It is sometimes difficult to accurately determine the groundwater level from the available data. Where uncertainty arises, a reasonably conservative estimate of groundwater level should be agreed between the Engineer and Contractor prior to commencing the works.

C1.11 Construction programme

A full day's notice to work outside normal hours may not always be possible (where, for example, it becomes clear at a late stage that safety or the integrity of the works could otherwise be compromised). The protocol for this scenario should be included in the Contractor's method statement.

C1.12 Records

It is important that adequate records are kept to ensure the smooth running of the contract and to enable, and demonstrate, control over the works. An over-emphasis on record keeping can, on the other hand, lead to attention being focused on the records rather than the object of the contract.

The particular records required by the Specification should reflect the requirements of the works undertaken. In some instances, additional records may be appropriate to verify, for instance, that design assumptions are valid. Similarly, in other circumstances, some of the specified records may be irrelevant and can be omitted without detriment to the contract. The specific requirements of the contract should be agreed before commencement of the works.

Pile completion or 'close-out' reports are a good way of documenting the experiences of the site team for later use. The rigour in producing these reports often helps with finalizing any outstanding quality issues. They are especially useful for situations where there is a need to re-examine the load capacities later, for instance, when some elements within the building have to become heavier and there is a need to verify if there is enough extra foundation capacity. They are also useful when the building is being later amended or its foundations are being re-used for a new structure. At that stage, many years after completion of the first building, most of the design and construction team are no longer contactable and there is no way of knowing whether particular piles were initially installed to their full capacity (or even installed to the specified lengths and diameters). The maintenance of records for piles is a requirement of EC7 (BS EN 1997-1 clause 7.9) and is also required under the CDM regulations.

It is important that the pile bore does not come to be regarded as a substitute for a thorough GI. Material from piling augers has suffered extensive disturbance and piling operatives do not commonly have a detailed knowledge of the principles of soil and rock description. Any data obtained from the pile bore should therefore be regarded as supplementary and be used only in combination with a thorough GI.

C1.13 Nuisance and damage
C1.13.1 Noise and disturbance

Guidance on the control of noise and vibration due to piling operations is given in BS 5228 and BS 7385.

C1.13.2 Damage to adjacent structures

The Engineer would normally determine the location and condition of adjacent structures and services that are likely to be within the zone of influence of the works. Certain structures may be particularly susceptible to noise, vibration or ground movement and the Engineer should assess prior to inviting tenders where special measures are likely to be required for protecting these structures. The information on location and condition of these structures together with the restrictions to be imposed and the monitoring requirements should be presented by the Engineer on the Drawings and in the Project Specification.

The Contractor is required to confirm the information on site and to provide proposals to meet the requirements of the Engineer in respect of these structures and services. If the Contractor believes additional structures are also at risk, he should bring them to the attention of the Engineer. As with the structures already identified, the Contractor should provide proposals to meet the requirements of the Engineer in respect of these structures and utilities.

Delays arising from late identification of services are often costly.

C1.13.3 Damage to completed piles or wall elements

When the concrete of recently cast bored, CFA or DA piles is still workable, vibration caused by the installation of adjacent piles may cause the ground to settle, or the loss of the concrete into loose material or into any voids. Also, in soft or loose ground the fluid concrete of nearby piles may be drawn into the bore, especially when the concrete pressure at any level in the bore exceeds the strength of the ground. These effects can cause a significant loss of section, or slumping of the pile shaft below cut-off level. Simple observation during construction should verify if recently cast piles are affected by slumping, or rising, of the concrete.

When the concrete of recently cast driven cast-in-situ piles is still workable, vibration caused by the driving of adjacent piles may cause the ground to settle, or loss of the concrete into loose material or into voids in fill. These effects can cause a significant loss of section, or slumping of the pile shaft below cut-off level.

When the concrete has gained an initial set but has little strength, any ground movement due to nearby pile installation may damage the pile.

Bored, CFA or DA concrete piles should not generally be installed within four diameters centre-to-centre of recently cast piles within 24 hours. In soft or loose ground, more than four diameters may be necessary. If piles are not filled with concrete but with a low-strength self-hardening slurry, then a larger clearance over a longer period may be necessary.

Driven concrete cast-in-situ piles should not generally be installed within six diameters centre-to-centre of recently cast piles within 24 hours. In soft or loose ground, more than six diameters may be necessary. Driven preformed piles may be installed closer than driven concrete cast-in-situ piles within 24 hours, subject to the requirements of Clause B1.14.

Should an excavation be made alongside completed piles, they will be subjected to lateral loading and the design should allow for such conditions where necessary.

Piles in soft or loose ground may be damaged by forces induced from plant loading. Piles should be designed accordingly.

Responsibility for protecting the works must implicitly be accompanied by the means to control the factors that could cause damage. The Contractor should ensure that all parties on site are aware of their responsibilities and the potential consequences of their actions.

In exercising his duties it is often unnecessarily restrictive for the Contractor to attempt to fully programme the sequence of pile installation. It is generally more appropriate to establish project specific guidelines covering the construction of adjacent piles.

Damage to completed piles can arise from a number of other factors including adjacent surcharges, vehicle impact and the effects of hydraulic breakers and cutters during pile trimming. A common cause of cracking in concrete piles is the use of machine mounted hydraulic breakers to trim them, particularly where the direction of impact is at an angle to the pile axis and especially with small diameter piles which have less resistance to bending and eccentric loading.

C1.14 Driving piles
C1.14.1 Driving procedures and redrive checks

The pile head should be adequately protected by packing to minimize head damage and transmit the impact force evenly over the pile head. The head packing should be checked periodically and changed as necessary.

The efficiency of a 'follower' is affected by its mass, fit and length. There is little published guidance in this subject. Comparative site tests may sometimes be possible. Where piles are driven to a set using a follower, the set calculation shall be adjusted to allow for the loss of efficiency within the follower.

The pile should be checked for verticality or rake before driving is commenced. Further checks for verticality or rake shall be made as driving proceeds. This is particularly important during the early stages of driving so that any appropriate corrective action may be taken. Normally corrective action can only be taken during the first 3 m of penetration. At greater depths such adjustments may overstress the pile. Positional tolerance shall also be checked by offset pins.

It is good practice, particularly in ground with obstructions, to conduct 'lift off tests' to periodically check the alignment of the drive assembly with the pile axis. In ground with obstructions (cobble-sized or larger) the Engineer should recognize that conventional tolerances may not be possible and account for this in the foundation design.

A driving record is required for every pile. As a minimum requirement, a complete driving record or blow count for the first pile in the works or section of the works shall be logged, including time of installation, time to join any elements and change of packing. Thereafter, driving records shall be taken at such intervals as site conditions and variability dictate. In uniform ground, detailed records for every pile may be unnecessary, with records for the final 1–3 m of every pile being adequate, combined with a full driving record of one pile per shift.

C1.14.3 Set

The Specification generally calls for the measurement of 'set' and 'temporary compression'. These generally need only be measured near to the founding level of the pile. For safety it is preferable that all measurements are taken remotely.

All parties should be made aware of the installation proposals and, in particular, as to whether the piles are to be driven to a set, a length or a penetration into a particular stratum.

Where the major part of the pile capacity is derived in shaft friction, the pile is normally driven to a penetration in the founding stratum. In some conditions driving resistance may be very misleading depending on the type of soil. High driving resistance, for instance, may arise from a build-up in pore water pressure and not due to the strength of the soil. Similarly low driving resistance can result from the mobilization of the soil immediately adjacent to the pile which can subsequently 'set-up' following the completion of driving. These conditions can be monitored by 'restrike' to check the set.

It should be routine to restrike at least one preformed pile per rig shift on the day following installation. This should be carried out

Specification for piling and embedded retaining walls. Thomas Telford, London, 2007

with the hammer operating in the same way as it was during the original driving to set. If the recorded set differs significantly from the original, a design and construction review will be required to define what action, if any, is required.

Where piles carry their load mainly in end bearing, the piles should be driven to a set in order to be satisfied that the pile is adequately seated. Sets calculated by driving formula may give misleading criteria and, wherever possible, the pile depth should be based on a static design and the penetration into the founding strata determined by taking into account variations in strength and relative density indicated by the SI and confirmed by the driving resistance achieved.

Noting the need for brief interruptions such as 'lift-off tests', the pile should be driven to its agreed depth and/or final resistance in a continuous operation to avoid 'set-up' occurring. If there is an unavoidable interruption to driving, it may be necessary to exceed the agreed driving criteria in order to found the pile at the approved depth established for nearby piles.

Pile sets can relax after installation due to ground conditions, or driving adjacent piles may cause heave, thereby reducing the overall performance of the pile in terms of load capacity and settlement. If this occurs then preformed piles can be redriven to their original set criteria if the pile positions are accessible for piling plant.

The timing of static and dynamic pile testing should take the effects of excess pore water pressure and pile set-up into account.

C1.14.4 Driving sequence and risen piles

The driving of piles inevitably causes lateral and vertical displacement of the ground. It may also lead to drift in pile installation making tolerances difficult to achieve. Since this can have detrimental effects on adjacent existing foundations, walls and piles, the Contractor should be given freedom to adopt a piling sequence which will minimize the risk of causing damage. If it is necessary to impose a less favourable sequence, the likely effects must be considered before proceeding with the works. The scale of deformations can be reduced by the use of pre-boring.

Where ground movement is likely to cause lateral movement of existing foundations or retaining walls it is normal practice to work away from the existing structures. In regard to previously installed piles, the piling sequence should be arranged to avoid driving large numbers of piles between previously installed pile groups. Similarly it is good practice to work outward from the centre of large pile groups. When installing raking piles any pile which rakes under another should be installed first.

Although a pile may be deviating from the specified tolerance in alignment and/or position, the Designer may decide, after any necessary design checks, that the pile may be incorporated into the works without affecting the integrity of the foundation, thus avoiding unnecessary delays and costs.

Where end bearing piles may have risen, the effects of heave may be reduced by pre-boring. Preformed piles may be redriven where this can be carried out without damage to the pile and the required access for the piling rig can be achieved.

C1.14.5 Pre-boring

Pre-boring is the removal or loosening (in non-cohesive strata) of soils, prior to pile driving, and may be appropriate in a number of situations:

- to ease pile driveability by breaking through soil layers where piles may prematurely refuse

- to reduce lateral soil displacement where this could cause damage to nearby structures (e.g. existing and newly installed piles, adjacent to retaining walls and services, and where work is carried within the confines of a cofferdam)
- to investigate and possibly deal with obstructions in the ground.

The loss of lateral restraint may allow the pile to buckle when driven into hard underlying materials and may make it difficult to achieve normally accepted piling tolerances. The diameter of the pre-bore shall be small enough to prevent buckling, or this shall be allowed for in the design of the piles. The pre-bore may also allow overburden to be dragged into the socket preventing full penetration.

If water jetting is used to ease driveability in granular soils, it must be used with caution. Jetting is carried out at high pressures and it may not be possible to maintain these pressures in open gravels. Conversely, in some soils this may lift adjacent footings. Jetting can also cause undermining of nearby foundations by removal of fines, particularly from gap-graded deposits.

C.1.15 Supervision and control of the Works

The level of site supervision provided should be sufficient to ensure the quality of the finished works, and must therefore reflect the scale and complexity of the works and the ground conditions. In extensive and/ or complex works the site supervision may be embodied in more than one person, resident on site. Conversely, where walling works are of a simple and minor nature forming part of a larger piling contract, a supervisor dedicated solely to the walling works would be a costly waste of resources.

C1.18 Definitions

Design Verification Load (DVL): particular conditions affecting the DVL include any factors which cause a difference between the loading conditions or the pile capacity which apply to the load test, and those applying to the working pile conditions.

Downdrag, which produces a time-dependent change in loading conditions, is caused by the settlement of soils relative to the completed pile shaft, and can result from a number of factors including:

- significant depths of recently placed fill causing ongoing settlements of the underlying soils or the fill itself. The SI should, if possible, identify the age of any fill since this affects the prediction of any likely future settlements
- floor loads from the proposed structure causing settlement of the underlying deposits
- ongoing secondary consolidation of soft alluvial deposits and organic soils
- reductions in water table resulting in consolidation settlement, although this may also improve the bearing capacity of piles in cohesionless soils
- the reversal of heave caused by installation of large displacement piles.

Differences in pile capacity commonly arise where the level at which load is applied during the load test is not the same as the permanent works cut-off level. Where the ground level is to be reduced by excavation after the piles are installed, the shaft length (and corresponding shaft friction/adhesion) are greater at the time of testing. This can either be negated by applying low-friction sleeving or slip-coating to the test pile, or allowed for by increasing the DVL. A similar situation pertains for downdrag, where the applied test load has to first overcome the positive skin friction from the ground plus mimic the negative skin friction that will occur in the long term. The practicalities of either approach should be agreed with the Contractor before finalizing the details. It may also be noted that changes in ground level or groundwater level will alter the allowable bearing capacity of piles founded in granular soils.

Factor of safety for a pile: the factor of safety need not be the same for different piling methods on the same site. The primary aim of a factor of safety is to ensure that settlements are within acceptable limits at Specified Working Load and do not increase disproportionately if the actual loads rise above the anticipated working conditions within a reasonable reserve of capacity. Where doubt exists, the adequacy of the factor of safety in controlling settlement may be verified by load testing.

King post wall: this is an economical form of wall construction compared to other methods, but the associated ground movements can be relatively large. This wall type cannot retain groundwater.

Project Specification: the Project Specification provides additional project specific information necessary to ensure that the completed works are constructed in an appropriate manner to meet the defined structural and functional requirements. This was termed the Particular Specification in earlier versions of *SPERW*.

Pile settlement: measured pile settlements may be affected by temperature fluctuations and bedding-in errors. Where possible these effects should be accounted for in their interpretation.

Ultimate Pile Capacity: Terzaghi identified the ultimate bearing capacity as 'the transition of the (load/settlement) curve into a vertical tangent'. The definition has been clarified in this document so that the Ultimate Pile Capacity is now defined as the maximum load which can be applied achieving the specified settlement rate criteria. Similarly this definition does not relate the ultimate pile capacity to a specific settlement criterion such as 10% of the pile diameter. The stated definition refers to a fundamental characteristic of the particular pile. In contrast, settlement related definitions refer only to random points on the load-settlement curve and tend to confuse conventional terminology. Acceptable serviceability states may be defined by various criteria, but to avoid confusion they should not be referred to as failure or ultimate limit states.

C2 Driven pre-cast concrete piles
C2.1 General

BS EN 12699 was published in 2001 and establishes general principles for the execution of displacement piles.

It is preferable that the design is based on pile sizes that are commonly available. The most common pre-cast segmental pile sizes are 187, 200, 235, 250, 270, 300, 320, 350 and 400 mm square. These are normally supplied in length increments of 1 m ranging from 2 m to 18 m overall segmental length. Lengths of up to 28 m are available for pre-stressed piles.

C2.3 Materials

There are several piling companies offering varying shapes and sizes of pre-cast concrete segmental piles, and different types of joint. Although the manufacturing process of individual companies may vary, whatever pile shape or size is chosen, the piles for a particular contract should comply with the requirements of the Specification.

Since a number of proprietary systems are available, the Engineer should request any relevant information pertaining to the system to verify the design. The use of non-standard pile sizes or reinforcement configurations will normally involve additional cost due to the implied adjustments to the manufacturing process. Where non-standard sections are required the Engineer should consider the effects on programme due to extended production times.

Although this clause refers to BS EN 12794, products complying with other equivalent international standards can also be considered.

Joints for segmental piles are generally made of steel and are therefore susceptible to corrosion if exposed to free oxygen and water. Consideration should be given to the location of the joints with regard to the risk of corrosion in completed piles. Joints in permeable ground, close to the surface of the water table, or in contaminated ground are at most risk. A problem with the final position of a joint may arise when pile driving reaches refusal prior to the intended depth of the pile.

C2.4.2 Handling, transportation and storage of piles

All piles within a stockpile should be in groups of similar length. Packing of uniform thickness should be provided between piles at the lifting points.

The stockpile platform should be level and competent to avoid damage which can result from bending and torsion in the piles.

C2.4.4 Strength of piles

Concrete cover for pre-cast driven concrete piles will be less than that used for other piling techniques but still sufficient for durability purposes as specified by BS EN 12794. Cover will be minimized so as to reduce the risk of concrete spalling off during hard driving.

C2.4.5 Pile installation system

During pitching and initial driving the pile may be guided at two levels, but it is common practice to only provide guidance at the pile head during driving. Should the pile be deflected during installation, the driving equipment should be adjusted so that the pile is subjected only to axial, non-eccentric blows from the hammer. This avoids the risk of damage due to eccentric loading of the pile.

The ground conditions may cause the piles to become driven outside the specified verticality tolerances. In this case a check will be needed to assess the impact on design.

C2.4.9 Cutting off pile heads

Heavy breakers can damage piles below the point of application. Either hand-held cutting equipment or hydraulic splitters should be used to trim the piles. When using hand-held cutting equipment, due consideration should be given to the effects of vibration on the operatives. In addition care needs to be exercised in excavating piles for trimming as mechanical excavators can easily damage slender pile shafts whilst digging.

A continuous connection between the pile head and the sub-structure is not necessary under purely compressive loading. It is common practice under continuous slabs to use a sliding joint to allow for thermal expansion of the slab.

C3 Bored cast-in-place piles
C3.1 General

This section applies to bored piles in which the pile bore is excavated by rotary and/or percussive means using augers, buckets, grabs or other boring tools to advance, where possible, a stable open hole. Where the bore is unstable, temporary or permanent casing or support fluids may be used to maintain the stability of the bore during excavation and concreting.

C3.2 Project Specification

Clause 2.3.4.2 of BS EN 1992-1-1 specifies that the diameter of cast-in-place piles for design has uncertainties and shall be allowed for in the design. Suggested reductions in diameter are given for circumstances where 'other provisions' have not been taken. In order to not reduce diameters for pile design, it is suggested that 'other provisions' could be the measurement and close control of auger and casing sizes on site.

It is preferable that the design is based on pile diameters that are commonly available. Construction of piles to special diameters requires manufacture of new tools and casings and may increase the costs of the project. Early discussions between the Designer and Contractor will help to establish availability. The more commonly available diameters range from 300 to 1200 mm in 150 mm increments and from 1200 to 2400 mm in 300 mm increments.

In practice the as-built diameter of a completed bored pile can be difficult to determine, particularly at commencing surface or in variable and coarse grained soils. It is therefore normal to base the pile diameter on the diameter of the casing and auger which are more readily accessible.

Tripod bored piles are rarely used in the UK, as it is safer to use a minipiling rig.

If minipiling techniques are to be used then the Project Specification should specify whether the acceptable pile installation tolerances are to be in accordance with BS EN 14199 or Table B1.4.

C3.3 Materials
C3.3.2 Casings

Whilst the uniformity of the internal section of temporary casings is essential to pile construction, externally it may be necessary to incorporate a slightly oversize driving shoe to assist penetration through dense/stiff soils. Care should be taken that this annulus does not create a void large enough to cause significant settlement/displacement at the ground surface, or allow groundwater to collect in the annulus, such that it is able to overcome the seal, and soften the shaft in the unlined bore. A protrusion of up to 10 mm may not produce a significant annulus in certain ground conditions.

Permanent casing may be used to:

- provide support to zones of the ground, or structures surrounding the pile which may become unstable or damaged before the pile concrete is set due to the pressures imposed by wet concrete
- provide additional structural capacity to the pile in the situation where the concrete plus reinforcement is unable to provide the required capacity, particularly the capacity to resist lateral loads
- provide a surface on which a friction reducing coating may be applied (see Section B16)
- provide a barrier against the ingress of contaminated or aggressive groundwater. Such casing must be in firm contact with the soils surrounding the pile to prevent the creation of a flow path. Any annulus should be filled
- form piles through voids or water, e.g. in docks.

Generally, any significant annulus outside the permanent casing should be filled with grout to prevent volume changes in the ground. The casing must be able to withstand the fluid pressure of the grout without buckling.

A range of casing types and treatments is available to cover the above varying requirements. In order for the Contractor to select an appropriate form of liner, it is important that the tender documents make the intended function of the casing clear to the Contractor. The information provided should include the purpose of the permanent casing, the number and location of piles which require them, the anticipated depths and diameters and the type and extent of any special treatments/coatings required (e.g. debonding agents).

C3.5 Construction processes

The more commonly occurring pile defects arising from construction processes are summarized in CIRIA report PG2.

C3.5.1 Placing of casings

Casings are generally installed into an open pre-bore that extends to a depth where instability is anticipated. The casing may be inserted into the ground ahead of the bore under its own weight combined with the rig and any crowding facility available in soft ground, or with a vibrator in more dense coarse grained deposits. In very hard ground (e.g. through obstructions and weak rock) the use of a powerful piling rig and/or an oscillator with segmental casings may be necessary although this is an expensive option.

In some situations, where vibration is not acceptable and the use of oscillators is not appropriate, casings may only be installed by mudding-in. This technique involves using the auger/boring tool to mix bentonite or clay with coarse grained soils, prior to excavation, to ease installation of temporary casings. The affected annulus of soil is generally no more than 50 mm thick around the pile. Such a thin annulus of material has little effect on the lateral resistance of the pile. The method is generally restricted to use in superficial deposits and it is common practice to either ignore or adopt a conservative estimate of shaft friction within these materials in the pile design.

The safety of all persons in the vicinity of open bores is of prime importance and should form an essential part of the method statements. The provision of a casing upstand 1 m above the working platform is standard practice. In some very limited circumstances this can make the handling of drilling equipment and reinforcement cages more hazardous. Also in low headroom there may be inadequate height to allow insertion of tools and reinforcement above a 1 m upstand. Similarly problems arise where lengthy tools such as underreaming buckets are to be used or where the piling rig is required to stand on a working platform lower than the pile commencing surface (e.g. at the foot of an embankment).

Where a 1 m upstand is not appropriate, a barrier should be placed around the bore to prevent accidents. Unattended open bores should be securely covered over to eliminate the risk of someone falling in. However, the use of a flat unidentified board/cover laid on the ground should be avoided since it masks the presence of the open pile bore. A sturdy sheet of mesh reinforcement may be a better solution.

C3.5.2.3 Placing concrete in dry borings

The length of discharge tube required should be such as to avoid segregation of the concrete as a result of it hitting the reinforcement or the pile walls. This may be particularly important in small diameter piles with heavy reinforcement.

C3.5.2.4 Placing concrete under water or support fluid

Where concrete to be placed under a fluid, a highly workable mix design should be selected and a tremie pipe must be used. During the pour, excessive depths of tremie penetration into the concrete can cause the concrete to stop flowing, leading to poor quality work. Equally, it is necessary to ensure that the tube remains embedded in the concrete throughout the pour to prevent segregation and avoid contamination.

Successful placing of concrete under a fluid therefore requires careful control, including accurate monitoring of the concrete and tremie levels, at each stage of the operation.

Where a tremie pipe is to be used it is important that the reinforcement is detailed to allow sufficient space for a tremie tube to be inserted and operated. BS EN 1536 gives recommendations regarding tremie dimensions.

C3.5.3 Stability of pile bore

The installation and removal of temporary casings becomes progressively more difficult and expensive with depth. This is particularly true for depths greater than, say, 12 m and where the casings are required to penetrate discrete layers of stiff fine-grained soil in order to stabilize underlying coarse grained deposits. In these circumstances the use of a support fluid may be a more appropriate solution.

Where a support fluid is used to provide stability it is necessary to ensure that a positive pressure balance, relative to the groundwater, is maintained within the bore throughout the drilling process. The insertion and removal of augers, even with care, causes displacement and disturbance of the fluid. The specified 2 m differential provides confidence that a positive pressure head is applied at all times. Where high water tables are anticipated, for example in the flood plain of a river, the use of support fluid may require that the working platform is raised in order that this differential can be provided.

The sudden loss of support fluid from the bore leads to instability and can arise from a number of factors:

- unidentified pipes and services (the most common cause)
- poorly backfilled Site Investigation holes
- open fissures and solution features in rock.

Where loss of support fluid does occur, it is inevitably an unexpected event and commonly requires immediate action to be taken to stabilize the ground and bore. The remedial actions to be carried out should therefore be agreed at the outset of the works to allow a rapid response by the Contractor. If the cause of the loss of fluid is such as to dictate a change in construction method, this should be agreed with the Designer. Any such change should be accompanied by a review of the design basis to ensure compatibility with the new method of construction. Where alternative construction methods are expensive or restricted, it may be necessary to revise the design.

C3.5.5 Continuity of construction

It is desirable that the time for which a pile bore is left open is restricted. The necessary time limit is, however, dependent on the particular ground conditions and the assumptions made in design. Some soils, such as sandy clays or clayey silts, may soften significantly in considerably less than 12 hours. On the other hand, bores into strong, stable rock may not be affected over several days.

It is sometimes not possible to complete a pile within the specified 12 hour period. This may be due to hard boring conditions and/or the required diameter/depth of pile, or may be due to a limitation on working hours. The extended construction period should be reflected in both the design and Specification.

Where prolonged delay in construction arises from unforeseen circumstances (e.g. unexpected hard boring/obstructions or plant breakdown) the pile bore may have to be backfilled to minimize deterioration of the shaft.

Any impact of the procedure adopted on the design should be considered by the Designer before the pile is completed.

C3.5.6 Underreams

The purpose of specifying a minimum height of concrete at the specified diameter of underream is to ensure that the design is not dependent on stress being applied at a 'feather edge'.

At the time of writing, top hinged underream tools are capable of meeting the requirements of the Specification if used correctly. Several contractors are currently developing new systems, so other systems may be capable of meeting the specification requirements in future. Untried methods should be evaluated against the Specification by full-scale testing prior to any use on a real contract.

C3.5.7 Cleanliness of pile bases
C3.5.7.1 General

The Designer must consider how much reliance it is necessary to place on the end-bearing element of the pile capacity and whether, as a result, particular measures to improve base cleaning should be specified. Special measures inevitably increase costs and reduce productivity. Measures such as inspection and cleaning of pile bases will increase the programme time.

In some ground conditions, inspection is not practical and attempts to improve the end-bearing performance may prove counter-productive. In sandy clays and clayey silts the additional time taken for base cleaning may allow the material beneath the base to soften, negating the advantage sought. Mudstones and chalks vary widely in character and can make it difficult to ensure that the base is clean. In other strata, such as water-laden silts and sands, it may not be at all practicable to obtain a sound, clean base. Under these conditions it may be preferable to consider an alternative method of construction.

Where there is uncertainty regarding the condition of the pile base, preliminary pile tests should be of sufficent extent to define the achievable end-bearing characteristics.

Alternatively, the pile design could incorporate a lower bound estimate of the end-bearing capacity in the working condition.

C3.5.7.2 Underreams

Due to the nature of underream piles (they carry a high proportion of load on the pile base) it is essential to ensure that the pile base is free from debris and remoulded material. The build up of remould on the underream base is highly dependent on the piling rig operator and, since underreams have rarely been constructed in the UK since the 1980s, there remain only a few UK operators experienced in underream construction. It is therefore likely that operators on any new underreaming jobs will have to follow a steep learning curve at the beginning of a project before they attain a consistently good standard of underream.

Clearly, rigorous inspection of each underream pile is required to satisfy the Engineer that the pile base is of good quality. Inspections have historically been carried out by manned descent of the pile; however, this has significant health and safety implications and is now not recommended.

Several UK piling contractors have now developed methods for inspecting piles using remote methods, combined with a high quality CCTV camera. It should be noted that CCTV inspection alone is not sufficient. Pile inspection must be supplemented by sampling and/ or probing to confirm the absence of remoulded material on the pile shelf.

C3.5.8.2 Underreams

The design of the system should be such that it can be guided positively by the CCTV in order to determine the orientation of the sample/test being taken/carried out.

The design of the sampler or penetrometer should be such that any foot required to provide a reaction against the underream shelf does not compress the soil around the tool in a manner which would improve the sample, or measured resistance.

Mist formed during boring can reduce visibility, making CCTV inspection impossible. The mist can be cleared more quickly by the introduction of an air line into the bottom of the bore. It is advisable to maintain an air line on site for this purpose.

C3.5.9.2 Concrete level

The construction of bored piles using temporary casings entails a number of uncertainties with regard to final concrete and reinforcement levels. Extraction of the casing from the pile bore allows the fluid concrete to flow to fill the remaining void and can cause movement of the unsupported reinforcement cage.

Predicting the magnitude of such movement is very difficult since it is affected by many factors including the casing dimensions and the compressibility, voids and stability of the surrounding deposits. The problem is accentuated in poor ground and where deep casings are required. In some circumstances it may be necessary to relax the specified tolerances.

C3.5.11.1 Grouting of piles

Post grouting of piles in the UK has most commonly taken the form of pressure grouting the pile base where the pile has been designed to take a significant end-bearing component.

The process is mainly used in sands where grouting is used to combat base disturbance and to ensure compatible performance of all the piles.

C3.5.11.2 Method of grouting

The Specification relates specifically to a base grouting system that uses V-tubes fitted with tube-a-manchettes. This is the system usually used in the UK.

There are, however, other systems of base grouting and also shaft grouting that have been used:

- Jacking plates — two steel plates sealed at the edge are placed at the pile base and grout is injected to force the plates apart to compress the soil beneath the pile base.
- Gravel basket — a wire basket filled with gravel is placed at the pile base and grout injected to fill the void spaces between the gravel and subsequently pressurize the base.
- Drilling and grouting — vertical steel tubes are installed in the pile close to the pile base with plugged ends. After concreting the pile a drill bores out the concrete below the tubes to the pile base and water is flushed from one pipe across to the opposite pipe to create a flow path for the subsequent grout injection.
- Post grouting of pile shafts has been carried out using tube-a-manchettes attached to the reinforcement cage down the side of the pile. This provides enhanced pile shaft capacity which is of benefit both for increasing the resistance to pile uplift during base grouting and increases the overall pile capacity.

C3.5.11.4 Pile uplift

The process of grouting must be carefully controlled to limit grout pressure, grout volumes and pile uplift within prescribed limits. Acceptable limits depend on the system and the soil conditions and should be left to the experience of the specialist contractor to define based on previous experience or site trials. The grouting installation and process require careful detailing and control to avoid potential problems. Some of these are highlighted below.

- Base contamination — where the pile is constructed under support fluid, concrete will be placed by tremie. Grouting pipework should be kept clear of the base of the tremie to allow free flow of concrete and congestion of pipework needs to be minimized to prevent trapping debris at the pile base.
- Pile slippage — delays in boring in a pile can lead to a build-up in the filter cake on the wall of the pile if bentonite support fluid has been used. This has been observed to result in uplift of the pile at very low grout pressures. If this is apparent, grouting should be stopped and the pile regrouted a minimum of 24 hours after the initial grouting.
- Hydrofracture — high grout pressures can cause soil layers to lift and separate resulting in large grout losses. This is indicated by a drop in grout pressure with a corresponding increase in grout take. By using slow rates of grout injection and viscous grouts of a paste consistency, this problem can usually be avoided.
- Uplift — uplift of the head of the pile can often be detected during grouting. For short large diameter piles, uplift is likely but no uplift may occur for longer thin piles.
- Stress relief — where piles are located within basements, stress relief due to removal of overburden may reduce the capacity of a grouted pile. Research indicates that the unloading effect is likely to be a function of changes in the mean stress rather than a change in vertical stress.

Pressure grouting is only appropriate in limited circumstances and requires specialist expertise in both design and construction. The technique should not be used to compensate for poor pile construction practice.

C4 Piles constructed using continuous flight augers or displacement augers
C4.1 General

This section applies to the following types of bored injected piles:

(a) Continuous flight auger piles (CFA) which employ a continuous auger for both advancing the bore and providing support to the surrounding soil. The spoil-laden auger is not removed from the ground until concrete is pumped into the pile bore through the base of the hollow-stemmed auger to replace the excavated soil. The reinforcement is inserted after the pile has been concreted to the surface. Typical CFA piles range in diameter from 0.3 m to 1.2 m in 0.15 m increments up to 25 m depth.

(b) Displacement auger piles (DA) which employ a displacement digging tool with following tube/auger which is screwed into the ground to the required depth thus creating a bore by displacing the surrounding soil. Spoil is neither flighted up the auger stem to ground level, nor collected on the auger for removal during the extraction process. Concrete is pumped into the pile bore through the base of the hollow-stemmed auger to replace the displaced soil. The reinforcement is inserted after the pile has been concreted to the surface. Typical DA piles can be either a continuous monolithic shaft with or without an enlarged base, or a continuous helical screw where the diameter of flights generated by the tool is larger then its structural core.

Permanent casing is not normally installed in conjunction with either of these pile types to any significant depth. This means there is no physical method of reducing friction over a significant length of pile.

C4.2 Project Specification

(a) Whether CFA or DA piles, or both, are acceptable.
(b) See Clause C4.4.6.
(c) See Clause C4.4.5.3.
(d) This item is intended to allow the specifier to request additional records in addition to the standard records required by the Specification.

C4.3 Materials
C4.3.1 Steel reinforcement

Longer cage lengths (greater than 8 m) are likely to require special measures, e.g. larger diameter bars to assist with handling and installation. Joints will normally be necessary where cage length exceeds 12 m. It is good practice to ensure that no more than one joint is present in a cage length. In low-headroom situations, where piles are being installed in accordance with Clause B4.4.6, it may not be possible to install the cages without more joints being used. In this situation, more than one joint may be allowed if permitted by the Project Specification, provided that the joints are made properly, and the cage inserted in a suitable manner.

Reinforcing cages should be designed and fabricated in such a manner as to withstand the effects of pushing/tapping the cage or the use of a vibrator.

Before specifying reinforcement debonding, or similar, discussions should take place with the Contractor. These will investigate if the proposal can be achieved for the chosen piling technique and diameters in the prevailing ground conditions.

C4.4 Construction processes
C4.4.1 Boring

During boring, when a CFA auger passes from a weak stratum to a strong one, there is a danger that the weak soils will be drawn up the continuous flight by a process known as 'flighting'. This produces local shaft enlargement and possible loss of integrity.

For CFA piles the drawing in of surrounding soils is most likely to occur in silts and fine sands, or when drilling from weaker soils into much stronger soils or rock. The risk is greatest in loose water bearing granular soils and is increased at larger diameters. It is best avoided by keeping the revolutions of the auger per metre penetration to a minimum.

Excessive penetration resistance or flighting in typical soils may be defined as a rate of penetration of less than 1 m per 10 auger revolutions for standard or heavy duty CFA augers and could be less than 1 m per 20 auger revolutions for extra heavy duty CFA augers. However, it is recommended that suitable limits should be determined for a particular site using trial bores or piles, and by agreeing suitable monitoring criteria with the Contractor.

It may not always be possible to advance the bore a further 0.5 m, for example, where the piles are founded on a hard stratum through which it is not possible to penetrate. Where the pile is dependent on the base capacity for design, then the rig shall be capable of achieving this additional depth but if the pile has already been bored to refusal then this additional 0.5 m will clearly be impossible to achieve.

If the pile shaft friction has been impaired, then this nominal extra 0.5 m penetration may not be sufficient.

It is good practice for the designer to have assessed prior to the Works a range of scenarios leading to the reduction of shaft friction so that contingency procedures for determining additional pile length or other requirements are available to the construction team.

C4.4.3 Removal of augers from the ground

It is a key requirement of CFA piling that the sides of the bore are never left unsupported. In unstable soils this is an essential aspect of the pile construction process. During boring the sides of the bore are supported by the auger loaded with spoil. During concreting the sides of the bore are supported by the fluid concrete. Even in stable soils, such as stiff overconsolidated clays, it is not good practice to construct the piles with any length of the bore unsupported.

For DA piles the following length of the auger may not have any flights and the bore will not then be supported by the tool over this length during boring. However, the construction of DA piles must ensure that the sides of the bore are never left unsupported during the concreting process.

If the base of the auger is not embedded in the fluid concrete during concreting, then the length of the bore below the base of the auger and above the head of the fluid concrete will not be supported. The use of instrumentation will help to keep the auger string embedded in the concrete over the pile length.

Concrete overflowing at the ground surface whilst the tip of the auger is still in the ground is a good indication that the pile is being concreted properly. If this is not observed then the concreting process should be reviewed. Topping up of the pile concrete once the tip of the auger is above ground may introduce inclusions, and is not recommended.

Spoil should be removed from the rising auger by automatic auger cleaners rather than by hand. This process should be carried out at a height such that spoil cannot spin off, endangering workers or members of the public. Particular care should be taken when working close to site boundaries or where rigid exclusions zones cannot be enforced.

C4.4.4 Depth of piles

The specified depth of piles could be depth penetration into rock, practical refusal or a founding level.

With CFA piles, the soil is not seen until concreting has commenced. With DA piles the soil is not seen at all. This makes it impossible to accurately assess penetration into a particular stratum especially if the superficial soils are of a similar strength or stiffness. It is thus advisable that pile depths are not specified in terms of penetration into a particular stratum (except penetration to practical refusal in rock) in the Project Specification. If piles are to be installed to practical refusal the designer should consider the power of the rig to be used.

If a pile fails to reach the required depth it should not be concreted without reference to the Designer. A number of options may be available:

(*a*) concrete the pile and review the design requirements afterwards (review design loads, group action, actual ground conditions, need for additional piles)
(*b*) leave auger in temporarily (risks jamming the auger)
(*c*) concrete and redrill pile (noting the requirements of C.4.4.4)
(*d*) reverse rotate the auger out from the pile and relocate the pile location (this would normally be a last resort).

C4.4.5 Placing concrete
C4.4.5.1 General

If cased CFA piles are used, concrete may not be to commencing surface level providing precautions are taken similar to those described in Section B3.

C4.4.5.2 Commencement of concrete supply to each pile

It is standard practice for the auger string to be raised slightly immediately prior to commencement of pumping concrete in order to assist blowing the temporary sealing device in the base of the auger. It is usual that the sealing device is removed by the application of concrete pressure before the base of the auger has been lifted.

Re-boring after the initiation of concrete flow is important in sands or silts or weak rocks where the suctions set up by lifting the auger may draw water and loose soil into the void. The action of re-boring is intended to ensure that a base of full cross-section has been formed and to help stabilize any ground instabilities that may have occurred. This could be mitigated by design by ensuring that the working load is less than the shaft capacity to ensure that no working load goes onto the pile base. This effect is most pronounced in larger diameter piles where the base provides a relatively larger proportion of the capacity, and it is particularly important to ensure that a competent base is formed. For smaller diameter piles this action may lead to concrete blockages, and should be carefully considered whether it is necessary to ensure that the full base capacity is reliable.

It is generally regarded as necessary for concrete to be significantly over-supplied at the start of the placing process in order for a pile toe of full section to be reliably formed. The technique is to raise the auger stem just enough to permit concrete flow. The base of the auger is then held steady to supply enough concrete to rise up the auger flights perhaps by 1 m or more before proceeding with the concreting process in the normal way. Where excess flighting has occurred, this may need to be much higher to fill the empty auger.

C4.4.5.3 Rate of supply of concrete

Rotation of the auger in CFA piles is necessary in many cases to ensure smooth auger lift and to clean the arising spoil from the flights. As in the boring process, the rate of rotation of the auger should be minimized during concreting. For some types of DA piles, reverse rotation of the auger is necessary in order to construct the pile of the given section characteristics, as it is necessary for the flight to return up the same void that it has created by being screwed into the ground.

There are two methods for controlling the placement of concrete. Method 1, which requires the Contractor to measure pressure, is the preferred method. Ideally pressure measurement should be at the base of the auger to measure the real values necessary to prevent defects occurring, but unfortunately it is usually measured above ground at the top of the auger. This is because Contractors have not been able to provide a pressure transducer at the base of the auger that can consistently withstand the aggressive forces which arise. It is also currently difficult to consistently transmit the data from the base of the auger through the auger string and its joints up to ground level. Consequently Method 1 is currently onerous, as the only concrete pressure measurements are the full auger height above the zone where ideally the measurement would be made. This leads to high concrete injection pressures which, while good for defect prevention, and probably better pile friction, are potentially wasteful in terms of the volume of concrete used. Accordingly, Contractors currently favour method 2.

When using method 1, it may not be practical to maintain a positive concrete pressure as the auger tip nears the ground surface because the applied concrete pressure may fail the surrounding soil or allow the concrete to flow preferentially up the auger flights. This will frequently occur, for instance, when the auger is embedded at 6–7 m depth in sands.

Method 2, which requires the Contractor to oversupply the pile concrete on a volumetric basis, is routinely used by Contractors at present. Concrete pressure should still be measured but need not be at the base of the auger. Concrete pressure will be measured at some point above ground, commonly at the highest point of the delivery tube (known as the 'swan neck'). If observed, a drop in concrete pressure tells the Contractor when the sealing device at the base of the auger has been successfully removed. If the concrete volume measuring system fails during concreting, the concrete pressure system may then be used to help complete construction of the pile. Negative concrete pressure measurements may be observed and this may be judged to be acceptable provided that they are isolated and not measured continuously throughout the pile length. These negative pressures are thought to be caused by piston effects within the auger and delivery tube and may not indicate that the pile integrity has been compromised. Negative pressures need to be considered in relation to the particular ground conditions to judge whether a problem has occurred.

The recorded concrete pressure is heavily dependent on the position of the transducer in the supply line and varies with parameters such as pump stroke volume. If measuring concrete pressure above ground, it is an indicator of relative pressure changes rather than an arbiter of the absolute pressure of concrete supply to the base of the auger. Positive pressure may indicate that the concrete is fully charged in the auger up to the transducer, but the pressure in the system can be influenced by external factors. Low-strength slurry mix or CFA piles constructed with grout rather than concrete may be topped up with low strength slurry mix or grout in accordance with the Contractor's Method Statement but only on the same day due to high shrinkage effects of these materials.

Method 1 is most suitable if there is a risk of voids in the ground or very soft layers are present, where over-supplied concrete may not contribute sufficiently to a sound concrete shaft. Method 2 is suitable where there is confidence that the whole volume of injected concrete is filling the pile bore and applying a net positive pressure against the surrounding soil.

Different concrete control methods are not permitted for preliminary and contract piles since, for method 1, the higher concrete pressures in the ground during pile formation may produce a pile which will perform better in a test, and incorporate fewer defects.

C4.4.5.4 Interruption in concrete supply

Piles are re-bored, for example, where a delay occurs during concreting.

C4.4.6 Splitting of augers

Where headroom is limited it may not be possible to construct bored injected piles without installing and removing the auger in sections. This technique may offer the only viable solution in some situations, and should only be considered when no other solution is available. Specifiers may more readily accept splitting of augers in ground conditions where the auger is not required to maintain the stability of the bore.

As the augers are joined during boring and split during concreting, it is difficult to provide a fully automated monitoring system for these piles. Accordingly, a manual system can be permitted provided it is capable of demonstrating that pile defects have not occurred during construction.

Careful control of the construction process, and in particular the concrete supply, will help to minimize the associated risks to acceptable levels.

This clause is not applicable when extending augers whilst working in areas of unrestricted headroom and using automated instrumentation systems.

C4.4.7 Placing of reinforcement

Typically bored injected piles have their reinforcement inserted on completion of concreting. The reinforcement is either pushed or vibrated into the concreted pile shaft/core. Difficulties may arise when inserting heavy or long reinforcement cages or where there has been a delay between concreting and cage insertion. Particular problems are experienced in unsaturated or dry permeable soils where the concrete stiffens rapidly or where the reinforcement cage has a number of links or joints that resist its penetration.

Reinforcement cages are installed after concreting to commencing surface level. Vertical tolerances on the level of reinforcement are dependent on the depth to cut-off level in relation to the commencing surface level as, in most instances, the top of reinforcing cages are left at commencing surface level.

The installation of DA piles causes ground heave which can induce tensile forces in the pile. Reinforcement for DA piles is therefore required to be installed to full depth in order to preserve the integrity of the pile.

Spacers can be provided to ensure correct alignment of the cage and thus achieve the minimum cover requirements. Cages that are inserted into workable concrete will tend to follow the track of the cage toe. It is therefore particularly important to guide the lower portion of the cage. This will in turn help to guide sections of the cage which penetrate any soft overlying soils where spacers will be less effective. Tapering of the bottoms of reinforcement cages can be introduced to ease cage installation.

Special measures may be needed where reinforcement is intended to penetrate more than 12 m into the concrete. Bars may, for instance, be bunched toward the bottom of the cage or a few centrally located bars may be used. Bars may also be centrally placed if the pile is not subject to bending. Additional vertical steel and welding may also be required to ensure that the integrity of the cage is maintained.

Any reinforcement debonding, or any other items attached directly to the pile reinforcement, can only be placed in the pile to the same tolerances as the pile reinforcement.

C4.4.9 Monitoring system for pile construction
C4.4.9.1 Automated monitoring system

The monitoring requirements for CFA piles are particularly onerous as CFA piles are the only pile type where the maintenance of pile bore stability is not observed, or observable. Experience shows that where monitoring of key parameters has not been undertaken, defects have occurred. This is also the reason why integrity testing of all CFA bearing piles is recommended.

The additional monitoring systems for DA piles will depend on the particular type of pile and its section characteristics. In most cases the system will be proprietary and the Contractor must demonstrate that monitoring system is capable of ensuring that the DA pile section characteristics are met.

The immediate availability of a hard or electronic copy of the monitoring output after completion of a pile can provide a common basis for discussion should an incident occur during pile construction.

Currently, piling rig torque is derived from measurement of hydraulic pressure.

C4.4.9.2 Manual monitoring system

Whilst the provision of a manual monitoring system is a sensible precaution, and allows the Engineer to independently check progress, it would only be used as the sole system in the event of failure of the automatic system whilst a pile was in the process of being constructed.

Pump stroke counters and flowmeters should be calibrated if used.

C4.4.9.3 Target boring and concreting parameters

The number of CFA auger revolutions which may be considered excessive during boring depends on the rig capacity, auger diameter, auger type, flight pitch, stem size, ground conditions and other factors. In general, however, the number of turns should be minimized and, for piles up to 600 mm diameter using standard or heavy duty CFA augers, the approximate maximum number of turns per metre might normally be expected to be less than 10. For piles 600 mm or larger, especially those using extra heavy duty augers, the maximum number of turns per metre may be in excess of 10, and could be as high as 30. The professional team may agree that the target maximum value need not be confirmed until after the construction of a number of trial bores, whose location should be chosen carefully. Caution needs to be given to underpowered piling rigs and to layered soils which have soft or loose layers overlying much stronger soils or rocks.

Target concrete oversupply varies with auger diameter and ground conditions. The minimum value is unlikely to be less than 5%. The maximum value is likely to be in the range 15 to 30%. For grout, higher values will be appropriate.

C4.4.9.4 Calibration

The CFA and DA pile construction process relies on the instrumentation to ensure the construction of sound piles. Hence calibration of the instrumentation at the start of piling and at appropriate intervals is essential.

Recalibration of the concrete volume will be necessary immediately if:

- the concrete discharge pump is changed or replaced as the pumps all have different characteristics. Same models of pumps will still need a recalibration.
- the length of concrete discharge tubing is changed as any calibration will only be appropriate for the length of tubing used at the time of calibration.
- the concrete mix is changed or is provided by a different supplier or batching plant, as any calibration is appropriate only for that concrete mix used at the time of calibration. A known value of concrete may be determined by the concrete delivery ticket.

C4.5 Records
C4.5.1 Records of piling

The records may be provided electronically or by hardcopy.

C5 Driven cast-in-place piles

C5.1 General

BS EN 12699 was published in 2001 and establishes general principles for the execution of displacement piles.

This section applies to piles in which a temporary casing is driven with an end-plate or plug, reinforcement placed within it and the pile formed by filling the temporary casing with concrete before, and during, casing extraction.

This section is also applicable to piles for which a permanent casing of steel or concrete is driven with an end-plate or plug, reinforcement placed within it as required and the casing filled with concrete.

Driven cast-in-place piles can either be top-driven (i.e. the hammer blows are applied to the pile head) or bottom-driven (where the blows are applied to an end-plate or concrete plug at the base of the casing). It is normal practice for piling contractors to offer a proprietary system for this pile type.

C5.3.1 Pile shoes

In most cases standard pile shoes can provide adequate watertightness during pile construction. However, under adverse conditions, it may be necessary to introduce a packer between the shoe upstand and the drive tube.

C5.4.1 Temporary casings

Temporary casings may use a heavier tube section or shoe at the base of the drive tube to withstand the driving stresses caused by obstructions in the ground and to reduce friction on the casing during driving.

C5.4.8 Placing concrete

Concrete mixes for cast-in-place piles are designed to be self-compacting. The application of vibration to such concrete is likely to cause segregation of the mix and is therefore not appropriate.

C5.4.12 Cutting off pile heads

Heavy breakers can damage piles below the point of application. Either hand-held cutting equipment or hydraulic splitters should be used to trim the piles. When using hand-held cutting equipment, due consideration should be given to the effects of vibration on the operatives. In addition, care needs to be exercised in excavating piles for trimming as mechanical excavators can easily damage slender pile shafts whilst digging.

A continuous connection between the pile head and the substructure is not necessary under purely compressive loading. It is common practice under continuous slabs to use a sliding joint to allow for thermal expansion of the slab.

C6 Steel bearing piles

C6.1 General

Detailed guidance on steel bearing piles can be obtained from the Steel Construction Institute (www.steel-sci.org) and the Steel Piling Group (www.webforum.com/steelpilinggroup).

C6.2 Project Specification

(*f*) *Grades of steel*
Steel bearing piles are normally manufactured to BS EN 10025 in grades S235, S275 and S355, but additional proprietary high-strength options are available from some manufacturers.

(*k*) *Pile shoes*
Toe protection shoes have been developed to prevent damage to the pile toe if it is penetrating debris, scree or other dense strata, the pile is being driven onto a sloping rock surface, or there is a need to increase the end bearing area of the pile. It is important to note that introduction of a shoe can lead to a loss of skin friction due to overcoring.

C6.3 Materials

Information on steel properties can be obtained from BS EN 1993-1-1 for bearing piles, BS EN 10210 or BS EN 10219 for tubular piles and BS EN 10248 for sheet piles and piles formed into box piles.

C6.3.2 Inspection and test certificates

In order to meet programme requirements it may not always be possible to inspect production of piles. In some cases the piles may be manufactured before contract award and be made available from stock. In such cases records of mill tests should be made available to the Engineer at least one week before work commences.

C6.4.1 Ordering of piles

It is important to note that steel rolling mills do not always operate on fixed manufacturing cycles. The period from order to supply may vary from a few weeks to a few months according to section type, total tonnage required and overall market demand. If steel sections require processing after rolling such as painting or fabrication this will add to the lead-in time.

Good communication and planning by all parties to the contract is essential if the programme requirements are to be met. Varying lengths, sizes, steel thicknesses or grades after an order has been placed can be expensive and may cause delay. There is a significant risk where piles are ordered before completion of the preliminary test piles.

C6.4.3 Handling and storage of piles

Treated piles should normally be stored on timber supports or plastic/hemp packing to protect the coating. Rigid piles can generally be lifted from the ends but flexible piles often require two or more lifting points or the use of a spreader beam.

Recommendations for lifting sheet piles are given in Annex A of BS EN 12063 and the same principles apply to lifting steel bearing piles.

C6.4.4 Driving of piles

Steel tubes can be driven open or closed as low or high displacement piles. They are usually top-driven by impact hammers or vibro-driving methods although pressing techniques have recently been developed for box piles. If piles are driven open, the soil on the inside may 'plug', i.e. the soil within the tube moves down with the pile. If the piles are to be driven closed, an end-plate is usually welded to the end with suitable stiffening plates to prevent distortion of the end of the pile. If the tube is driven open and the soil inside is subsequently excavated, considerable care is required not to induce inflows of water or soil into the tube.

H-piles and hollow sections are essentially low displacement piles as the volume of material driven into the ground is low. However, if the section plugs with soil during driving to the extent that the plug moves down with the pile, the soil displacement will increase and the means by which resistance is generated will change from predominantly skin friction to end bearing. However, the steel section can be amended by the addition of plates and wings to enhance performance. Plating across the base of the section will increase the cross-sectional area of the pile which may be advantageous if the ground to be piled is suitable for generating end-bearing resistance rather than skin friction. Similarly, attaching short lengths of the same section at appropriate positions along the pile shaft can be used to increase the surface area of the pile in strata which can provide high skin friction resistance.

In situations where piles need to be driven into a sloping rock surface it is possible to prevent the pile toe slipping down the slope during driving by welding a steel bar or pin to the toe of the pile which will dig into the rock surface and prevent movement. Having a smaller cross-sectional area than the pile, the pin concentrates the driving force onto a small area of rock causing it to penetrate the rock surface and hold the pile in place. The pin will generally be formed using high-strength steel to ensure that it will not buckle under the driving force.

It is possible to form a box pile by combining varying numbers of sheet-pile sections and, by using pile connectors to join the sections together, it is possible to use pressing techniques to install these piles as each pile can be driven independent of its neighbours. This is advantageous in urban areas as the noise and vibration normally associated with the installation of welded-box piles can be eliminated and the piles can be extracted at the end of their useful life.

C6.4.11 Extraction

It is recommended that the Contractor uses equipment which can provide an extraction force of at least twice the ultimate shaft friction of the pile.

C6.5 Coating piles for protection against corrosion

Considerable guidance is available on design corrosion rates for different situations within Section 4 of BS EN 1993-5 and the 8th edition of the *Piling Handbook*. The corrosion rate applicable to steel piles installed into undisturbed natural soils tends to zero as both water and oxygen must be present for corrosion to take place. At depth there is little oxygen available so the process ceases.

C6.5.8 Thickness, number and colour of coats

Refer to Part 7 of BS EN ISO 12944.

C6.6.1 Site welding

Magnetization of the heads of driven steel piles can occur and may result in magnetic arc blow during butt-welding of pile extension pieces, with detrimental effects on the quality of the weld. Degaussing of the pile head is necessary in these circumstances and can be achieved by the generation of a counteractive magnetic field during welding.

C6.6.2.1 Welded tubular piles

Where the use of uncertified welded piles is acceptable, appropriate testing (both destructive and non-destructive) should be carried out as required by the Designer.

C7 Timber Piles
C7.1 General

Timber piles for permanent structures should only be used below the lowest anticipated groundwater table or free water level during the lifetime of the structure unless adequate protection is provided.

Further guidance on the design and installation of timber piles, suitable species and preservative treatment is given in BRE Digest 479 and in BS EN 12699.

C7.2 Project Specification

The first criterion to establish prior to installation of timber piles is what their end use is to be and what tolerances are required for the end product. For example, orientation in plan can be more critical if the pile is to be used to form a continuous structure such as groynes used for sea defences. Position and verticality may be more critical if the pile is to be used as a load-bearing member of the structure. The overall appearance of the piles relative to each other and/or the remainder of the structure may be more important where piles are to simply be used as an architectural feature.

In groyne or wall applications, the position and orientation of the pile relative to the next pile is often more critical than the absolute position of the wall itself. As tides can also affect the pile position by releasing stresses in the ground, continuous monitoring is needed during installation to note any pile deviations so that they can be taken into account in the relative positions of subsequent piles. Final pile positions should therefore be checked and agreed immediately after pile installation. It may be necessary to place gates on the ground and/or attached to the piling rig to assist with accurate positioning of the piles. For other applications the amount or type of driving control should reflect the end requirement.

C7.3 Materials

Piling is a demanding application for timber, requiring material which is:

- strong and straight-grained
- adequately durable (or able to be made so with preservatives)
- available in large cross-section.

In addition, the general rise in concern for environmental issues requires timber certification on many projects.

C7.3.1 Species/grade of timber

Relatively few species comply with all three of the structural criteria, the most commonly used being:

- greenheart (a tropical hardwood)
- ekki (a tropical hardwood)
- oak (a temperate hardwood)
- Douglas fir (a softwood).

The structural grades given in the specification for tropical hardwoods and softwoods are appropriate for piling work. For temperate hardwoods, the two grades given in BS 5756 (THA and THB) are widely separated in quality. For the lower grade (THB) a slope of grain of 1 in 4 is allowed, which may be considered too permissive. For the higher grade the knot ratio is limited to 0.2 of the faces on which they occur which may make it difficult to find timber of the necessary quality. Enquiries should be made with suppliers.

C7.3.4 Preservatives

The natural durability of timber heartwood varies between species, while the sapwood of all species is effectively non-durable. Five durability classes are defined in BS EN 350-1, ranging from 1 (very durable) to 5 (not durable). The natural durability, and treatability with preservatives, of common species is given in BS EN 350-2.

BS 8417 gives recommendations for the durability class of timber in various locations. Timber in fresh water should be class 2 or better; timber in salt water should be class 1. Note also the relatively short life-expectancy of piles in salt water. This recommendation can be complied with in one of two ways:

- selecting a species with appropriate natural durability
- treating the timber with preservative to achieve the required durability.

Of the timbers suitable for piling, most hardwoods are extremely resistant to preservatives, and the softwoods are not durable. Thus the choice is generally between:

- hardwoods of durability class 2 or better, or
- softwoods treated with preservatives.

Consideration should be given to using the timber in the round (i.e. not sawn). The band of sapwood on the outside of round timber is normally more permeable than the heartwood and can be penetrated with a high loading of preservative. This can provide a better degree of protection than that to be expected from the limited preservative penetration often obtained in the heartwood of timber classed as resistant or extremely resistant.

There are two preservatives which can be applied to timber piles; creosote and copper-based systems.

The Creosote (Prohibition on Use and Marketing) (No. 2) Regulations 2003, which implement a European Directive, allow creosote to be used for the treatment of timber piles. The Environmental Protection (Controls on Dangerous Substances) Regulations 2003, SI 2003/3274 and The Marketing and Use of Dangerous Substances (No. 4) Regulations (Northern Ireland) 2003, SR 2003/548 allow the use of CCA-treated (copper-chrome-arsenic) timber for constructional use in fresh water and brackish water, but not in marine waters.

The use of creosote is heavily restricted so it is advisable to check current guidelines with the British Wood Preserving and Damp Proofing Association (www. bwpda.co.uk) and the Environment Agency.

In the UK there are still many creosote treatment plants, but very few processors currently offer CCA treatment, and it should not be specified despite its inclusion in some current standards. The alternative copper-based treatments now offered are not yet the subject of a Standard and enquiries with processors is advised to develop a Project Specification.

Guidance on wood preservatives is given in the Standards noted above and BS 1282, and further information and more details can be obtained from the British Wood Preserving and Damp-proofing Association and the Timber Research and Development Association (www.trada.co.uk).

C7.3.5 Certification of timber

The two basic aims of certification are:

- legality
- sustainability.

The first of these is to provide assurance that the supplier or their agent was entitled to harvest the timber under the laws of the country of origin, and a clause to this effect should always be included in any specification.

The second requirement is not yet the subject of a universally agreed definition. It can be used in the limited sense that the supply comes from a 'sustained yield' source, one that is managed to provide a continuous supply of timber. In the interpretation of environmental groups it covers much more than this, for instance, including the maintenance of biological diversity, without damage to forest eco-systems or forest-dependent people.

The UK government has made a commitment on all central government departments and agencies actively to seek to buy timber from legal and sustainable sources, and this should be borne in mind for projects where the government or its agents are the client. It is up to other clients to decide on the degree to which they wish to follow the same path.

Since timber is commonly used at some distance from the original area of growth, a number of internationally recognized forest certification schemes are now in operation, including:

- the Canadian Standards Association (CSA)
- the Forest Stewardship Council (FSC)
- the Malaysian Timber Certification Council (MTCC)
- the Programme for the Endorsement of Forest Certification (PEFC)
- the Sustainable Forest Initiative (SFI)

and suppliers may be able to provide timber which is certificated under one of the above schemes. It should be borne in mind that certificated timber still forms a minor part of the total market and delivery times may be extended. At a less demanding level, many UK suppliers have adopted an Environmental Purchasing Policy, and can often confirm that all timber supplied has been obtained from legal sources, and that it has been harvested in accordance with the laws and regulations governing forest management in the produced country.

It is, however, important to issue a specification which does not contain requirements that are in practice unrealistic. It is strongly recommended that enquiries are made with suppliers so that the perceived importance of certification, or other assurances, is balanced against the species properties.

Further information on certification can be obtained from the Governments Central Point of Expertise for Timber Procurement (www.proforest.net/cpet) and from the websites of the certification organizations noted above.

C7.3.6 Pile shoes

Spikes and points tend to induce splitting and capping rings tend to come off during driving. A flat plate is preferred to provide protection to the pile end and to prevent splitting.

C7.4 Construction processes

If spikes or points are fitted to the pile and are not square, it may not be possible to drive the pile vertically as it will naturally tend to deviate off line. Capping rings should be fitted tightly and perpendicular to the long side of the pile to reduce any tendency to come off during driving and a flat plate may be preferred to provide protection to the pile end and to prevent splitting. Piles should be driven with a packing at the head and the ends should be chamfered if no fittings are to be used.

C8 Diaphragm walls and barrettes
C8.1 General

This section applies to diaphragm walls and barrettes in which a trench (the excavation) is formed either by grabs or reverse circulation mills. Grabs using either rope or hydraulically operated clam shells advance the open excavation by removing material in separate bites while reverse circulation mills allow almost continuous removal of material within the support fluid returns.

Support fluid is used to support the walls of the trench prior to concreting. Each completed element is known as a panel. Since each panel is concreted individually it is necessary in the case of walls to form the joint between panels either using stop-ends or by overcutting joints using reverse circulation mill equipment.

C8.2 Project Specification

(*f*) Preliminary barrettes are barrettes which are constructed and load tested in advance of the construction of the working barrettes to verify the construction methods and design performance.

(*g*) Using some stop-end systems it is possible to incorporate a continuous water stop at each panel joint. A properly designed and installed water stop will inhibit, but not eliminate, water flow through the joint. Experience has shown that a single water stop per joint is as effective as two or more water stops.

C8.3 Materials
C8.3.1 Support fluid

Section C18 gives detailed information on support fluids.

Table 1 of BS EN 1538 gives characteristics for bentonite support fluid. This table is based on the various compliance values used throughout the European Union for bentonite support fluids from several different sources in a wide variety of ground conditions. It is reproduced below as Table C8.1.

C8.3.4 Alternatives to steel reinforcement

To facilitate easier breaking through of diaphragm walls at positions of tunnels and services, the Contractor may propose to use alternative materials to steel reinforcement, such as glass fibre or polyester bars. Special precautions will be necessary for handling and lifting reinforcement cages comprising these materials.

C8.4 Construction tolerances
C8.4.1 Guide wall

The top edge of the finished internal face of the front guide wall, closest to any subsequent excavation, represents the reference line for the diaphragm wall. The face of the wall should not contain

Table C8.1 Bentonite properties from Table 1 in BS EN 1538:1996

Property	Stages		
	Fresh	For re-use	Before concreting
Density (g/ml)	<1.10	<1.25	<1.15
Marsh value (s)	32–50	32–60	32–50
Fluid loss (ml)	<30	<50	n/a
PH	7–11	7–12	n/a
Sand content (%)	n/a	n/a	<4
Filter cake (mm)	<3	<6	n/a

Notes:
1. The Marsh value is the time required for 946 ml to flow through the orifice of the cone.
2. The fluid loss is the value obtained in the 30 minute test.
3. 'n/a' means not applicable.
4. At the stage 'before concreting', an upper limit value between 4% and 6% for sand content may be used in special cases (e.g. non-load bearing walls, unreinforced walls).

Table C8.2 Diaphragm wall verticality tolerances with special control measures

Panel excavation method	Achievable verticality tolerance
Cable grab	1:100
Hydraulic grab	1:150
Reverse circulation mill	1:250

ridges or abrupt changes. The rear guide wall should be constructed to similar tolerances to the front guide wall.

C8.4.2 Diaphragm wall and barrettes

The accuracy with which the guide walls are constructed will determine the position of the top of the panel. During excavation, the excavation tool can deviate from the vertical and can also twist, and both need to be closely observed and controlled as far as possible. A vertical tolerance of 1:75 can normally be achieved with all types of excavation tool, and this is probably sufficiently accurate for many diaphragm walls.

Higher tolerances may be achievable with special control measures, see Table C8.2, and if required for structural performance or to maximize basement space these will then need to be specified in the Project Specification. Such tolerances may be needed for construction purposes, for example to facilitate the removal of temporary stop-ends or for the placement of irregularly shaped reinforcement cages.

Concrete protrusions on the exposed face of the wall can be caused by overbreak during excavation. A tolerance of 100 mm is normally sufficient for protrusions when the wall is constructed through naturally occurring granular soils, firm clays without boulders, or weak rocks which can be excavated without the use of a chisel. Through very soft clay, peat and made ground, or where obstructions or boulders have to be removed, or where chiselling is necessary, overbreak well in excess of 100 mm can occur. It is not possible to predict with any certainty how much additional overbreak will occur in any of these circumstances, and this should be recognized in the Project Specification. Additional overbreak may also occur at the internal corners of tee panels and corner panels, and may be unavoidable in some types of soil.

C8.4.3 Recesses

Box-outs for recesses should be avoided if possible, because they restrict the flow of the concrete and can trap supporting fluid. If they cannot be avoided, their dimensions should be kept to a minimum, and they should be tapered at the bottom and the top to reduce the risk of supporting fluid becoming trapped when the concrete is placed. They should be used with caution near the top of a wall where there may not be sufficient concrete above the box-out to ensure sound completion of the concreting operation. Box-outs should not extend behind the outer layer of reinforcement and should not extend beyond the ends of the reinforcement cage into the unreinforced sections of wall at the panel joints. The material used to form box-outs must be securely fixed to the reinforcement cage to ensure that it is not displaced by the concrete.

Bars which are required to project from the face of the wall to connect into floor slabs can be provided in one of three ways.

1. By casting bars into the wall and bending them out later. This method should be restricted to steel reinforcement up to 20 mm diameter. Using this method for the ground bearing slab has resulted in joint performance difficulties and is generally not recommended.

2. By casting bars with couplers into the wall. The couplers can then be exposed on the face of the wall and connecting bars screwed into them.
3. By drilling into the wall after it has been exposed and fixing bars in position by grouting.

The latter method is preferable to the two previous methods because it does not require transverse bars to be cast into the wall, and the projecting bars can be located with greater accuracy. It is the only method which can be used if bars are required at tremie pipe positions. However, consideration will need to be given to the risks associated with hand/arm vibrations.

Projecting bars cannot be provided between the reinforcement cages at panel joint positions by casting bars into the wall, and drilling into the wall too close to the joints may cause leakage. Gaps must therefore be left in the slab connecting reinforcement at these locations.

Perforations through diaphragm walls for anchors or services should preferably be formed with tubes not exceeding 300 mm in diameter, with sufficient space between the tubes to allow the concrete to flow freely around them. Larger diameter tubes may impede the flow of the concrete and should be used with caution. Rectangular perforations will almost certainly trap supporting fluid within the concrete, and should not be used.

C8.4.4 Reinforcement

It is important to ensure that each panel is excavated within the specified tolerances so that the reinforcement cage can be inserted with sufficient concrete cover on the front and rear faces and also at the ends. It is particularly important when excavating tee panels, corner panels and the like to ensure that the excavation is carried out as accurately as possible. It is also important that additional concrete cover should be allowed to the reinforcement in such panels to ensure that the cage will fit into the excavation after taking into account allowable tolerances. Taking a 20 m deep tee panel, for example, the leg of the tee and the top of the tee could be 250 mm from their true positions at the base of the excavation and still be within the specified tolerance. One or both of them could also be twisted. A reinforcement cage with only the minimum allowable cover would not fit into this excavation.

Congested reinforcement and box-outs for recesses in diaphragm wall panels can lead to defective concrete. The concrete must be able to flow easily around the bars and the box-outs, and rise uniformly over the entire area of the panel. Typical details are shown in Fig. C8.1.

When deciding on the diameter and spacing of the vertical bars, it is important to remember that in the case of wall panels, the reinforcement will not extend to the ends of the panel, because space has to be allowed between adjacent cages for the panel joint (see Fig. C8.2), end cover to the reinforcement cage and allowable construction tolerances. The reinforcement should not extend into the shoulders of the panel joints since this will obstruct concrete flow at a particularly critical location. The unreinforced section may represent about 10% of a typical panel length but could be up to 25% of the length of a single bite panel. The final clear distance between the bars will therefore be less than it would otherwise be if the bars could be spaced uniformly along the length of the wall.

The final clear horizontal distance between vertical bars in a single layer of reinforcement should be at least 100 mm. This figure can be

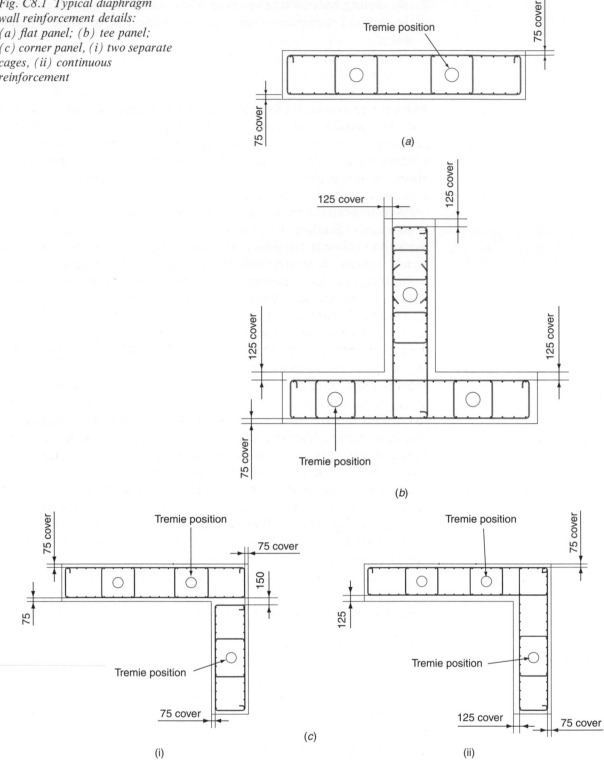

Fig. C8.1 Typical diaphragm
wall reinforcement details:
(a) flat panel; (b) tee panel;
(c) corner panel, (i) two separate
cages, (ii) continuous
reinforcement

reduced to 80 mm over lap lengths. Where two layers of reinforcement
are required on a wall face, the bars in the inner layer should be
aligned behind those in the outer layer in order to allow the concrete
to flow easily between them.

Fig. C8.2 Typical diaphragm
wall joint details: (a) tubular
stop-end; (b) tubular with flange;
(c) proprietary stop-ends

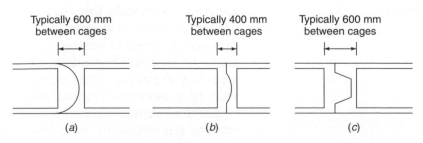

Specification for piling and embedded retaining walls. Thomas Telford, London, 2007

The clear vertical distance between horizontal bars should be a minimum of 150 mm provided a maximum aggregate size of 20 mm is used. Otherwise the minimum clear distance should be 200 mm.

Links are used between the front and rear faces of a reinforcement cage to hold the bars in their correct positions and to provide shear resistance. In vertically spanning panels the links should enclose the vertical bars to satisfy design requirements and to provide lateral restraint to the vertical bars when the cage is being lifted. The clear horizontal distance between the legs of links should be at least 150 mm, with at least 500 mm at tremie pipe positions (see Fig. C8.1). The same clear distances are required between transverse bars at connections between floor slabs and the diaphragm wall.

Joints in steel reinforcement can be made either by lapping the bars or by using couplers. Cages should be made in one length where possible in order to minimize the delay between completion of panel excavation and commencement of concreting. If a cage is too long to be lifted in one piece, it must be made in two or more pieces which are then joined together as they are lowered into the panel excavation. Joining the sections of cage together over the excavation will be facilitated if the cage is fabricated in one piece on the ground then split into sections before being lifted. This is of particular advantage when couplers are used.

Diagonal bracing bars are normally provided in each section of cage to ensure that it will not distort when being lifted. Specially designed lifting bars are provided at the top of each section, to which lifting shackles can be attached, and specially designed hanging bars are provided at the top of the upper section to enable the cage to be suspended at the correct level in the excavation.

If there is insufficient space available on site to fabricate the cages, it may be possible to make them at a separate location and transport them to site by road. In these circumstances the width of the cages has to be restricted to the maximum that can be transported, therefore two or three separate cages may be required in each panel. Multiple cages may also be necessary in the case of panels founded on hard rock where a varying founding level is required. The minimum clear distance between adjacent cages in the same panel should be 200 mm.

C8.4.5 Concrete casting level

There will be some intermixing of the concrete and the support fluid at their interface, and there may also be a layer of thick contaminated support fluid on top of the concrete. It is important to ensure that all this material is lifted up above the cut-off level and that all the concrete below the cut-off level is sound. In order to achieve this, the concrete should be brought up to 1 m above the cut-off level. Where the cut-off level is less than 1 m below the top of the guide walls, it will be necessary to flush some of the contaminated concrete over the top of the guide walls to ensure that the concrete below the cut-off level is sound.

C8.4.6 Dimensions of panels

Panels can be constructed in various shapes on plan, but the most common are straight panels, tee panels, and corner panels. Standard panel widths are 600, 800, 1000, 1200 and 1500 mm. Panel lengths depend on several factors including the dimensions of the excavation tool, the type of stop-end equipment (if required), the excavation sequence, the ability of the support fluid to maintain trench stability and the proximity of adjacent structures. The volume of concrete required to fill the panel and the weight of the reinforcement cage are also important considerations.

The length of straight panels can vary from a minimum of about 2200 to 3000 mm, equal to a single bite of the excavation tool, to a maximum of about 5500 to 7000 mm, equal to approximately 2.5 bites of the excavation tool. Close to adjacent structures, panel lengths are often restricted to a single bite in order to increase trench stability and reduce ground movement effects.

Tee panels reinforced with a single cage are used to support the sides of deep excavations where their shape provides additional strength and stiffness. The lengths of the top of the tee and the leg of the tee, behind the wall, are usually restricted to a maximum of approximately 4500 mm in order to keep the volume of concrete and the weight of the reinforcement cage within practical limits.

The internal angle of a corner panel is usually 90° but could be greater or smaller in an irregular shaped excavation. Each leg of the panel can consist of a single bite of the excavation tool, or one leg may consist of a single bite and the other may be equal to approximately 1.5 bites. It is generally necessary to reinforce such panels with L-shaped cages to provide continuous reinforcement to ensure structural continuity, but it is important to note that this may not always be practical, particularly in the case of deep panels where individual cages may be necessary.

Shafts are sometimes constructed using a series of corner panels but, in this case, the internal angle between the legs of the panel is much greater than 90° and depends on the diameter of the shaft and the number of panels. Alternatively, each panel may consist of three bites but this increases the volume of concrete per panel and the weight of the reinforcement cage. The use of reverse circulation mills together with overcut joints can enable the construction of shafts exceeding 50 m in depth without the need for stop-ends.

Owing to the variations in the type of equipment used, giving rise to preferred panel lengths and working sequence, the Contractor should be allowed to determine the panel lengths and working sequence, subject to any limitations specified in the Project Specification.

C8.4.7 Water retention

Diaphragm walls, when subjected to significant water pressure, are rarely watertight. The achievable degree of watertightness is influenced by several factors including:

- piezometric pressure in relation to the final excavation level
- permeability of the retained soils
- wall movement during excavation.

Even without wall movements, small gaps (say 1–2 mm) will unavoidably result from shrinkage as concrete temperatures fall.

Visibly flowing water can be treated by grouting but damp patches are likely to remain. There is a lack of available data on which to base realistic estimates of leakage per individual square metre of wall but total water ingress is unlikely to be significant after removal of visible flows.

Waterproofing of the connection between a diaphragm wall and a basement slab requires careful consideration. Proprietary systems are available for sealing the joint between the face of the diaphragm wall and the basement slab, but water can still flow up the front of the joint from permeable strata below formation level, and around the back of the sealing system in some cases. The normal method of overcoming this problem is to inject grout into the diaphragm wall joint after the basement slab has been constructed.

C8.5 Construction processes

C8.5.1 Drawings

Construction drawings are an essential part of the planning, procurement, fabrication and construction process. Drawings can often be very detailed and can only be undertaken upon completion of the final design. A realistic period for approval of the drawings should be agreed between all parties at tender stage.

C8.5.2 Guide walls

Guide walls are necessary to achieve the required positional tolerances. They act as a guide for the excavation tool as it enters the ground, and can be used to monitor verticality as excavation proceeds by observing the position of the suspension ropes or kelly bar. The ropes or kelly bar should remain midway between the guide walls if the excavating tool is hanging vertically.

Guide walls also provide support for the reinforcement cages before and during concreting, and for jacking equipment when stop-ends are extracted vertically in the construction of some diaphragm walls. Guidance frames for reverse circulation mills are often clamped inside guide walls. Finally guide walls act to provide containment for support fluid required to ensure trench stability.

Since the guide walls provide a reference line for the construction of diaphragm walls and barrettes, it is important that they do not move. They are usually constructed in reinforced concrete, the reinforcement being in the form of steel mesh with additional longitudinal bars to provide continuity across the construction joints. Temporary struts are provided in the trench between the walls to maintain the correct separation before panels are excavated, and excavated material is placed between the walls to prevent the flow of support fluid along the trench.

In stable ground conditions, each guide wall is usually about 1 m deep and 300 mm thick. The depth of the walls may have to be increased through made ground where there could be loss of supporting fluid. Alternatively, the made ground may have to be removed completely over the width of the guide walls plus at least 1 m on each side, and replaced by well-compacted material which will remain stable below the guide walls when the diaphragm wall is being excavated.

In soft ground, or where heavy vertical loads are to be applied to the guide walls, it may be necessary to provide horizontal projections at the base of the walls to reduce the bearing pressure on the soil and provide additional resistance to horizontal pressures induced in the ground by the heavy excavation equipment. In extreme cases, ground improvement may be required to ensure guide wall stability.

In circumstances where there is a high water table, and it is not possible to lower the water level or to raise the ground level to the required height above the water table for trench stability, the guide walls can be raised above ground level to maintain the required level of the support fluid. This option, however, can impede plant movements across the line of the wall and introduce difficulties in concreting and stop-end removal, and should be avoided, if possible.

C8.5.3 Stability of the excavation

Care must be taken during the excavation process to ensure that the guide walls are not undermined. The level of the support fluid should be kept as close as possible to the top of the guide walls, and sudden variations in level due to rapid extraction of the excavation tool should be avoided. In the case of panels left open overnight or over a period of days, measures should be put in place to ensure that the support level is topped up as necessary.

The level of the support fluid would not normally be less than 1.5 m above the level of external standing groundwater at all times.

Particular attention should be paid to trench stability in the case of panel excavation where groundwater levels may be tidally influenced and to the specific case of the closure panels of a shaft where internal lowering of the groundwater level may be required.

In the case of diaphragm wall construction, the sequence of construction depends on the method of joint formation to be used and the number of excavation rigs to be employed. In order to avoid congestion on the site, the rigs should be spaced as far apart as possible, and each rig should have two areas in which to work.

In the case of walls constructed using stop-ends, at the commencement of operations a starter panel is constructed in each area. These are followed by intermediate panels, and finally closure panels which are constructed where the intermediate panels of each area meet. The intermediate panels may be constructed sequentially or in a 'hit and miss' pattern, depending on the type of stop-end equipment being used. Each excavation rig will alternate between its designated working areas in order to allow space for fluid cleaning operations, reinforcement cage installation and concreting operations after excavation of each panel has been completed.

C8.5.4 Cleanliness of the base

In the case of barrettes and diaphragm wall panels where a significant proportion of vertical load is to be carried in end bearing, the base of the excavation should be cleaned out prior to concreting. In this regard the method of cleaning, which may utilize a combination of grab, submersible pump and/or air lifting techniques, should be determined to take into account the panel geometry, excavation method, ground conditions and type of support fluid.

C8.5.6 Stop-ends in diaphragm wall panels

Temporary stop-ends are used to form the joints in the concrete between panels. They are not designed to prevent water and soil from entering the excavations. The shape of the joint formed in the concrete by the stop-end varies between contractors. Some contractors use circular stop-ends in conjunction with grabs that have circular-shaped jaws. Others use rectangular or keyed stop-ends in conjunction with grabs that have square-shaped jaws, or with rotary cutters. Some stop-ends are extracted vertically, before the concrete has finally set. Others are extracted by releasing them horizontally, after the concrete has set, by pulling sideways into the next panel excavation. Some stop-ends will remain permanently within the wall. Water-stops can be incorporated into the joints in some cases. The joint detail therefore depends on the type of equipment that the Contractor proposes to use.

Experience has shown that the watertightness of overcut joints is poor, except in the particular case of circular shafts where hoop compression is believed to help control leakage through the joints.

C8.5.7 Placing concrete

In order to provide a cohesive concrete mix with the required consistence for placing through a tremie pipe, it is normally necessary for the sand content of the concrete to be more than 40%, by weight, of the total aggregate. The target slump should be 200 mm (or its equivalent flow table value) and concrete with a slump of less than 150 mm should be rejected.

It should be noted that BS 8500 allows a tolerance on specified slump which may permit a specified slump that would be unacceptable for diaphragm walling purposes. Discussion and agreement of the slump requirements with the concrete supplier is essential before commencing the works. The use of target flow values to specify workability is more appropriate for high slump mixes.

Concrete must be placed beneath the supporting fluid through one or more tremie pipes. Tremie pipes normally have a nominal internal diameter of 250 or 300 mm and an external diameter of approximately 300 or 350 mm, measured at the joints. The number of tremie pipes required in a panel depends on the shape of the panel and the horizontal distance over which the concrete has to travel. It is good practice to limit this distance to approximately 1.8 m, in which case a single tremie pipe can be used in a 3.6 m long panel. Two tremie pipes will be required in panels between 3.6 and 7.2 m long, and three tremie pipes will usually be required in a tee panel.

During concreting, the tremie pipes must remain immersed in the fresh concrete. It is recommended that the minimum embedment below the surface of the concrete should be 3 m, but this can be reduced to 2 m when the level of the concrete is accurately known. The immersion depth may have to be reduced when the concrete approaches ground level to facilitate the necessary flow. Excessive embedment of the tremie pipes in the concrete can cause the concrete to stop flowing or may cause the reinforcement cage to rise. The maximum embedment should therefore be limited to 6 m.

C8.6.2 Records of special control measures during excavation

Such records may include rig instrumentation records.

C9 Secant pile walls
C9.1 General

This section applies to secant pile walls whether constructed by continuous flight auger or casing techniques.

It is preferable that the design is based upon diameters that are commonly available. Construction of piles to special diameters requires the manufacture of new tools and will increase the costs of the piling. Early discussion between the Designer and the Contractor will help to establish availability. Primary piles may be constructed using a different diameter to the secondary piles.

C9.2 Project Specification

Integrity testing of secant pile walls by impulse or sonic echo methods is not likely to give meaningful results because the signal will be affected by adjacent piles.

C9.3 Materials
C9.3.1 Self-hardening slurry mixes and low-strength concrete mixes

Self-hardening slurry mixes

In the short term, shrinkage and cracking has not been found to be significant. Little data exists to demonstrate long term durability which will be affected by the mix proportions, air temperature, humidity, temperature gradient and soil conditions. These mixes used in hard/soft pile walls are therefore generally regarded as providing only short-term water retention.

Accreditation by NAMAS is unlikely to cover tests on these mixes, which exhibit different characteristics to soils and to concrete. These mixes require different handling and testing procedures and those associated with slurry walls are more appropriate.

The ICE *Specification for Slurry Walls* provides guidance for the requirements for self hardening slurries.

Low-strength concrete mixes

These mixes are used in hard/firm and hard/hard pile walls. The early age strength gain of the concrete in the primary piles is critical to the secant pile construction. It is therefore preferable not to specify a 28 day concrete strength for primary piles. It is preferable to use a prescribed mix with a high cement replacement content to achieve the required early age strength gain characteristic.

Such high cement replacement contents mean that there is often a conflict between the immediate needs of the Contractor and the long term durability requirements of the Designer. When the primary pile is required by the Designer to fulfil a long-term function as part of the structure, then a compromise will need to be sought between the Contractor and the Designer. This compromise must form part of an integrated risk management strategy for the structure, and will differ from project to project.

C9.3.2 Support fluid

For secant pile walls, when full-length casing is used for bored cast in place piles, it may be found that water provides a suitable fluid to maintain the stability of the pile base.

C9.4 Construction tolerances
C9.4.2 Secant piles

The Designer should specify only those verticality tolerances necessary to achieve the desired quality for the end product. Unnecessarily restrictive tolerances inevitably lead to increased costs. The Project Specification can be used to specify tolerances which are more onerous than those given in Table B1.4. Typical tolerances which can be achieved for each construction method are given in Table C9.1.

A tight control on verticality will be essential where rigid structural elements, such as steel I-sections, are to be inserted into the piles, particularly where there is little clearance between the edge of the section and the side of the bore. In cases where the reinforcement of primary

Table C9.1 Secant pile wall verticality tolerances

Piling method	Verticality	Depth of pile interlock
Continuous flight auger (CFA) using standard or heavy duty augers	1:75	0–5 m
CFA using extra heavy duty augers*	1:125 1:75	0–7 m 7–10 m
Cased CFA	1:150	0–15 m
Bored cast-in-place using standard tools	1:100	0–7 m
Bored cast-in-place using stiffened casings with cutting teeth	1:200	0–20 m

* Augers comprising thickened stems

piles is required, very careful detailing of cages and/or sizing of I-beams is essential. This should take into account both verticality tolerances of pile and cage/I-beam. Under such circumstances, early consultation between Designer and Contractor is strongly recommended.

It is very important to match the piling technique with its associated tolerances to the structural requirements. Here Table C9.1 can be used as guidance.

In favourable ground conditions the tolerances are achievable with existing technology. However, in some ground conditions, or where there are obstructions in the ground, the required tolerances may not be practical. The Designer should discuss with the Contractor early in the design process to assess the viability of the design.

For secant pile walls constructed with continuous flight augers, experience shows that the piles do not deviate from vertical linearly with depth, but follow a curve. Thus Table C9.1 gives a varying tolerance with depth to take account of this effect.

C9.4.3 Recesses

Structural connections and recesses should be constructed after completion of the pile since it is not possible to consistently place them accurately. In addition, box-outs also impede concrete flow and can result in zones of poorly compacted or contaminated concrete within the pile.

If structural connections and recesses are required to be incorporated within piles then the Designer should discuss with the Contractor early in the design process to assess viability and the alternative use of diaphragm walling techniques should be considered.

C9.4.4 Reinforcement

While the reinforcement details must meet the structural design criteria, it is vital that they are appropriate to the piling construction method. The Designer should consult with the Contractor as necessary to produce these reinforcement details.

Particular attention is required to ensure that bars are not so closely spaced or arranged that they prevent proper flow and compaction of the concrete. A clear bar spacing of 100 mm should be regarded as the minimum requirement for vertical bars and 175 mm for horizontal bars. If piles are constructed with support fluid then larger clear bar spacings are encouraged.

Multiple layers of reinforcement are to be avoided as they further impede concrete flow. When increased structural capacity or flexural stiffness is required, this is more reliably achieved by increasing the pile diameter.

Where a minimal clear bar spacing is specified, it is essential that the Contractor maintains strict control of the concrete mix proportions, the concreting process and the properties of any support fluid.

To avoid lifting of reinforcement during casing removal, or to assist plunging reinforcement into continuous flight auger piles, it may be necessary to increase cover above 75 mm. Where full-length casing is required, a minimum clear distance of 50 mm between the inside of the temporary casing and outer cage diameter should be maintained. Cover may be reduced below 75 mm for primary piles, if reinforced.

To help avoid lifting of reinforcement during casing removal, the Contractor may choose to rest the reinforcement cage on the base of the pile.

C9.4.5 Water retention

The achievable degree of watertightness is influenced by several factors including:

- piezometric water pressure relative to the final excavation level
- soil permeability
- depth of excavation in relation to verticality tolerances achieved
- number of joints
- wall movements during excavation
- number of changes in direction in the line of the wall
- differential vertical or horizontal loading.

In common with diaphragm walls, secant pile walls, when subject to significant water pressures, are rarely found to be perfectly watertight. The greater frequency of joints in secant pile walls will make them more vulnerable to water ingress.

Whilst visibly flowing water can be treated, damp patches are likely to remain. It is difficult to obtain relevant data on which to base realistic estimates of leakage through a secant pile wall. Also it is not possible to accurately measure inflow and then to determine whether the inflow is attributable to the piling subcontractor or to any follow on works.

Where the secant pile wall forms part of the permanent works, it is good design practice to provide appropriate drainage and ventilation to the excavated face. BS 8102 and CIRIA 139 provided guidance.

Hard/soft pile walls are generally regarded as providing water retention in the short term only. Where long-term water retention is required, a secondary lining system is required to provide the water retention. Alternatively, the use of a hard/firm or hard/hard secant pile wall, or a diaphragm wall, may be appropriate.

C9.5 Construction processes
C9.5.1 Guide walls

Scalloped guide walls are necessary to achieve the required construction tolerances. These walls provide restraint to the tool position and, to some degree, verticality. They also offer support for reinforcement cages, casing extractors, etc. during the concrete pour.

Straight guide walls offer no restraint to the tool position along the line of the wall and thus tolerances in this direction will be greater.

The guide wall level should be set above the top of the pile reinforcement in all cases.

Since the guide walls provide a reference line for pile positioning, it is important that they do not move prior to or during pile construction. Guide wall design should allow for continuous reinforcement and for lateral propping. In poor ground, regular checks on guide wall position will be necessary to ensure that they have not moved in any direction.

C9.5.2 Bored cast-in-place piles

For bored cast-in-place piles, bore stability may be achieved using casings. Subject to achieving the necessary verticality, the Contractor may choose to use an oversize casing over the depth of unstable soils during construction of the primary pile. This action will have the effect of increasing the size of the pile cut, but only over this depth. However, this will negate the requirement for a full-length casing for the primary pile. Alternatively, continuous flight auger techniques may be adopted.

In the case of free-standing walls without any applied vertical loading, the cleanliness of the pile base may not be of primary importance. The Contractor may choose not to apply any special measures to improve base cleanliness in this instance. Where vertical load is applied to the pile wall it may be necessary to place reliance on the end bearing capacity of the piles. Special measures to improve base cleanliness would then be necessary.

C9.5.4 Boring near or into recently cast piles

The boring of the secondary pile involves cutting into the primary pile. It is therefore important to recognize the time window that the Contractor has to bore the secondary piles successfully. The longer the primary pile is left to harden, the more difficult it will be to bore the secondary piles to the required depth and with the required tolerance.

The construction sequence should be arranged so that the secondary piles are bored when the adjacent primary piles have sufficient strength to resist damage, but are not too strong to prevent the joint being cut. The time to cut the joint should not be less than 24 hours from casting the last primary pile at the joint, and ideally within four days. Where a clear difference in pile age is foreseen, say at the beginning of a run, then a dummy pile (i.e. bored but not concreted) should be constructed at the secondary pile position.

The Contractor may choose to install casing within adjacent piles over the depth of unstable soils, prior to concreting, thus avoiding any possibility of concrete flow from pile to pile.

C10 Contiguous pile walls

C10.1 General

This section applies to contiguous pile walls whether constructed by continuous flight auger or bored cast-in-place techniques.

It is preferable that the design is based upon diameters that are commonly available. Construction of piles to special diameters requires the manufacture of new tools and will increase the costs of the piling. Early discussion between the Designer and the Contractor will help to establish availability.

C10.2 Project Specification

Integrity testing of contiguous pile walls by impulse or sonic echo methods is not likely to give meaningful results because the signal will be affected by adjacent piles.

C10.3 Materials
C10.3.1 Support fluid

For contiguous pile walls, when full-length casing is used for bored cast-in-place piles, it may be found that water provides a suitable fluid to maintain the stability of the pile base.

C10.4 Construction tolerances
C10.4.1 Guide walls

The construction of contiguous pile walls using guide walls is uncommon. However, the utilization of guide walls, if properly designed and constructed, will improve construction tolerances and should be specified if plan positional tolerances better than ±75 mm are required and/or ground conditions are particularly unfavourable.

C10.4.2 Contiguous piles

The Designer should specify only those verticality tolerances necessary to achieve the desired quality for the end product. Unnecessarily restrictive tolerances inevitably lead to increased costs. The Project Specification can be used to specify tolerances which are more onerous than those given in Table B1.4.

A tight control on verticality will be essential where rigid structural elements, such as steel I-sections, are to be inserted into the piles, particularly where there is little clearance between the edge of the section and the side of the bore.

It is very important to match the piling technique with its associated tolerances to the structural requirements. Here Table C10.1 can be used as guidance.

C10.4.3 Recesses

Structural connections and recesses should be constructed after completion of the pile since it is not possible to consistently place them accurately. In addition, box-outs also impede concrete flow and can result in zones of poorly compacted or contaminated concrete within the pile.

If structural connections and recesses are required to be incorporated within piles then the Designer should discuss with the Contractor early in the design process to assess viability, and the alternative use of diaphragm walling techniques should be considered.

Table C10.1 Contiguous pile wall verticality tolerances

Piling method	Verticality
Continuous flight auger (CFA) using standard or heavy duty augers	1:75
CFA using extra heavy duty augers[*]	1:125
Cased CFA	1:150
Bored cast-in-place using standard tools	1:100
Bored cast-in-place using stiffened casings with cutting teeth	1:200

[*] Augers comprising thickened stems

C10.4.4 Reinforcement

While the reinforcement details must meet the structural design criteria, it is vital that they are appropriate to the piling construction method. The Designer should consult with the Contractor as necessary to produce these reinforcement details.

Particular attention is required to ensure that bars are not so closely spaced or arranged that they prevent proper flow and compaction of the concrete. A clear bar spacing of 100 mm should be regarded as the minimum requirement for vertical bars and 175 mm for horizontal bars. If piles are constructed with support fluid then larger clear bar spacings are encouraged.

Multiple layers of reinforcement are to be avoided as they further impede concrete flow. When increased structural capacity or flexural stiffness is required, this is more reliably achieved by increasing the pile diameter.

Where a minimal clear bar spacing is specified, it is essential that the Contractor maintains strict control of the concrete mix proportions, the concreting process and the properties of any support fluid.

To avoid lifting of reinforcement during casing removal, or to assist plunging reinforcement into continuous flight auger piles, it may be necessary to increase cover above 75 mm. Where full-length casing is required, a minimum clear distance of 50 mm between the inside of the temporary casing and outer cage diameter should be maintained. Cover may be reduced below 75 mm.

To help avoid lifting of reinforcement during casing removal the Contractor may choose to rest the reinforcement cage on the base of the pile.

C10.5 Construction processes

C10.5.1 Guide walls

The use of guide walls should be specified where the required construction tolerances or ground conditions dictate such measures. Guide walls which fully define the circular shape of each pile may be necessary to achieve the required construction tolerances. These walls provide restraint to the tool position and, to some degree, verticality. They also offer support for reinforcement cages, casing extractors, etc. during the concrete pour.

Straight guide walls offer no restraint to the tool position along the line of the wall and thus tolerances in this direction will be greater.

The guide wall level should be set above the top of the pile reinforcement in all cases.

Since the guide walls provide a reference line for pile positioning, it is important that they do not move prior to or during pile construction. Guide wall design should allow for continuous reinforcement and for lateral propping. In poor ground, regular checks on guide wall position will be necessary to ensure that they have not moved in any direction.

C10.5.2 Bored cast-in-place piles

For bored cast-in-place piles, bore stability may be achieved using casings. Alternatively, continuous flight auger techniques may be adopted.

In the case of free-standing walls without any applied vertical loading the cleanliness of the pile base may not be of primary importance. The Contractor may choose not to apply any special measures to improve base cleanliness in this instance. Where vertical load is applied to the pile wall, it may be necessary to place reliance on the end bearing capacity of the piles. Special measures to improve base cleanliness would then be necessary.

C11 King post walls
C11.1 General

King post walls are defined in Section B1.18. This section applies to king post walls inserted into pile bores whether constructed by continuous flight auger or bored cast-in-place techniques. The horizontal units, whether they comprise timber, steel or pre-cast concrete planking or sprayed concrete, and which are installed to retain the earth between the king posts during bulk excavation, will be referred to as lagging in this section.

Such walls are generally constructed for temporary works purposes and therefore designed by the Contractor. Since they are not watertight they are only appropriate for retaining water-bearing soils if temporary dewatering or other special measures are taken to enable excavation.

C11.3 Materials
C11.3.2 Support fluid

Ground conditions, which require the use of a support fluid to construct the pile, are likely to preclude the use of a king post wall. Unless special measures are taken, installation of the lagging requires locally stable and reasonably dry soils. In cases where the groundwater table lies below excavation level, but dictates the use of support fluid for pile construction, the additional costs are likely to make other forms of wall more effective.

C11.4 Construction tolerances
C11.4.1 Piles

King post tolerances should be compatible with the subsequent form of lagging. Less onerous tolerances will be required for cast-in-situ concrete lagging.

When specifying tolerances it should be borne in mind that unless the pile diameter is significantly larger than the maximum (diagonal) width of the I-section, the accuracy which can be achieved is restricted by the tolerances applicable to piling. Where tighter tolerances are to be specified, either larger diameter piles will be required with special measures to allow accurate placing of the king post within the bore or more accurate piling techniques will be required.

In both cases, costs will be increased and other forms of wall may be more appropriate.

C11.4.2 King posts

The section alone normally provides the structural capacity of a king post wall. The primary purpose of the concrete in the lower portion of the shaft is to provide fixity, transferring loads by horizontal compression into the ground. It is therefore generally not appropriate to provide a reinforcement cage in the concrete. An exception to this case might be where it is required to extend the embedded portion of the shaft beyond the bottom of the king post to provide additional fixity.

Where king post walls are only constructed for temporary works purposes, issues of long-term durability are not relevant. It would only be necessary to consider incorporating reinforcement to meet serviceability requirements if the king post wall was required for long-term purposes. Such an application would be unusual.

If steel reinforcement is used below the section, then the tolerances should be stated in the Project Specification.

The rotational tolerance specified should be compatible with the design and/or all other construction tolerances.

C11.5 Construction processes
C11.5.3 Placing king posts

The method of placing king posts will depend on the method of construction of both the pile and the subsequent lagging. If it is necessary to place concrete by tremie under water or support fluid, it is preferable that the king post is inserted after completion of concreting since concreting using twin tremie pipes can be problematic.

In dry bores the king post can be placed either before or after concreting.

Where preformed lagging is to be inserted between posts it is necessary to achieve greater accuracy in placing the king posts than where cast-in-situ panels are to be constructed.

C12 Steel sheet piles
C12.1 General

This section covers interlocking steel sheet pile sections which are generally constructed to resist horizontal soil and water pressures and/or superimposed loads. It is not relevant to piles constructed from timber, plastic or any other material apart from steel. Sheet piles are used in both permanent and temporary situations, e.g. cofferdams, retaining walls, etc. and are usually installed using impact hammers, vibrators or hydraulic pressing. The piles have interlocking clutches which connect together to form a continuous retaining structure. In the case of sheet piles the Contractor, particularly in the case of temporary works, commonly undertakes the design.

Although this section refers to standard interlocking sheet piles, it is also relevant to 'combined walls' i.e. walls consisting of differing primary and secondary elements — commonly tubular, H or high modulus sections combined with sheet piles.

Sections of Z-type steel sheet piling which have their interlocks in the flanges develop the full section modulus of an undivided wall of piling under most conditions. Sections of trough or U-type steel sheet piling with close-fitting interlocks along the centre line or neutral axis of the sheeting develop the strength of the combined section only when the piling is acting compositely. The shear forces in the interlocks may be considered as resisted by friction due to the pressure at the walings and the restraint exercised by the ground. In certain conditions it is advisable to connect together the inner and outer piles in each pair by welding or crimping to ensure that the interlock common to the pair can develop the necessary shear resistance.

Cold formed sections which comply with BS EN 10249 are considered in this section. It should be noted that the sheets may not interlock tightly and are prone to separation if piles are installed inaccurately or in difficult soils. Cold formed sheets may have a lighter section range than hot rolled piles and are more suited to less heavily loaded walls in easy ground conditions.

C12.3 Materials and fabrication

Steel sheet piles are one of the most demanding sections to roll as a result of the tight tolerances needed to ensure that sections will interlock. However, dimensional tolerances for sheet piles are not generally as strict as those that apply to other forms of structural steelwork. This reflects the fact that it is not normally necessary to install sheet pile walls with the same accuracy as structural steelwork. Tolerances on the dimensions of sheet piles are covered by BS EN 10248 for hot rolled piles and BS EN 10249 for cold formed piles. Material obtained from sources which do not manufacture to these Standards may be subject to variable dimensional tolerances, particularly those of the pile clutch and the web thickness.

Rolled or extruded corner sections are available as alternatives to prefabricated corners as are rolled or extruded junction sections. These are intended for fabrication on site so that lead times are minimized and accurate positioning of corners may be achieved to counter wall length creep.

When long sheet piles are to be driven, there is an increased likelihood of declutching and it may be necessary to consider specifying the interaction factor of the clutch in the design.

C12.3.2 Fabricated sheet piles

Details for fabricated sheet piles are included within BS EN 12063 and examples are given in the 8th edition of the *Piling Handbook*.

C12.3.5 Clutch sealant and seal welding

Design guidance about sheet pile wall permeability and watertightness can be found in Annex E of BS EN 12063, Chapter 2 of the 8th edition

---

of the *Arcelor Piling Handbook*, and the Arcelor publications entitled *The Impervious Steel Sheet Pile Wall*, Parts 1 and 2.

C12.4.1 Ordering of piles

The period from order to supply of steel piling by the manufacturer can vary significantly according to the type of section, total tonnage required and demand. If the steel sections are also to be coated, this will add to the lead-in period. Good communication and planning by all parties to the contract is essential if the programme requirements are to be met.

Manufacturers will require the following information in order to provide an accurate quotation for both cost and anticipated delivery period:

(*a*) Section type/name
(*b*) Interlocked form
(*c*) Crimping arrangements (if required)
(*d*) Section length(s)
(*e*) Steel grade
(*f*) Tonnage
(*g*) Coating requirements (if required)
(*h*) Lifting hole requirements (if required)
(*i*) Sealant requirements (if required)
(*j*) Corners and junction type (if required).

C12.4.3 Handling and storage of piles

Guidance regarding storage and handling of steel sheet piles is given in Annex A of BS EN 12063 and the 8th edition of the *Piling Handbook*.

Sheet piles should be stored on timber supports or similar to prevent distortion of the flanges and interlocks. Rigid piles can generally be lifted from the ends but flexible piles often require two or more lifting points or the use of a spreader beam. Particular care is needed when handling and storing long piles or flat, straight web piles. Soft web slings may be used to move bundles of coated piles to avoid paint damage.

Sheet piles of the U-type are normally supplied and pitched singly or in crimped pairs.

Sheet piles of the Z-type are normally supplied and pitched in crimped pairs to make installation easier and quicker although heavier equipment may be needed. Crimping prevents elements sliding differentially and keeps the flanges of the piles in line, thus avoiding rotation.

The Contractor may opt to supply un-crimped pairs for temporary works purposes in order to extract the piles singly or when using hydraulic pressing equipment where rams drive the individual pile elements separately.

Recommendations for lifting piles are also given in Annex A of BS EN 12063.

C12.4.4 Pile installation

The Designer must consider the driveability of the piles. Further guidance is provided in Annex C and D of BS EN 12063, the 8th edition of the Arcelor *Piling Handbook* and the NASSPA installation guide. Sheet pile sections should be selected to ensure that they are capable of withstanding the bending moments and axial loads which will be applied during and after construction. They must be capable of being driven through the soil to the required penetration. Any limitations or restrictions on methods of assisting driving such as water-jetting or pre-augering must be made clear in the Project Specification.

Toe pinning can be adapted to work with sheet piles when they are to be driven onto a sloping rock surface or where rock exists at high level and there is insufficient depth to develop the required passive resistance in the soils above the rock. In this case, toe pins can be used to develop additional resistance to movement by requiring them to shear before the pile toe can move.

Sheet piles may be installed using panel driving and/or pitch and drive techniques. Panel driving involves the erection of a temporary guide frame into which a panel of piles is pitched and interlocked prior to being driven. It also requires a crane which has sufficient reach to be capable of interlocking a pile with one already standing in the vertical position. As pile interlocking occurs at height with the panel driving method, a safe working platform is essential and care is required to ensure that the pile is under full control at all times. The use of pile threaders and ground release shackles is recommended as these preclude the need to work at height.

The pitch and drive method of installation will normally involve a leader rig which supports one pile unit — which may be a single pile or pair of piles — as it is interlocked and driven partially or full length into the ground. It is essential that piles installed using this method can be driven with adequate vertical control. It is recommended that a rigid ground level waling is used to prevent excessive twisting of the piles by the leader rig during the driving and correction process. If excessive lean occurs piles should be withdrawn and corrected before continuation using panel driving techniques.

Panel driving is the preferred method of installation for overcoming obstructions and driving in hard soils. Initially the degree of lead between each pile toe may be several metres but as driving becomes harder the advance needs to be suitably limited to prevent declutching.

Where noise and vibration are of concern, vibrationless jacking systems are available for installing sheet piles. Some proprietary systems can only be used in stiff cohesive soils whereas others can push piles through cohesionless soils although they may be limited in the penetrations they achieve.

A system for ground pre-treatment is available for use with pile pressing methods. It is possible for an auger to loosen the soil as the pile is being driven. The process is controlled as there is a finite distance between the toe of the pile and the position of the auger.

The selection of the correct driving equipment requires experience of piling in similar ground conditions. Mathematical analysis of the driving process may also be used, as an aid to confirm selection of the pile section and size of impact hammer required.

| *C12.4.9 Methods of assisting pile installation* | Any method which assists pile installation must be compatible with the design assumptions regarding the state of the ground around the pile after installation. Proposals to adopt such methods require approval of the Engineer prior to commencement so that the design is not compromised. |

C12.5 Coating piles for protection against corrosion

Guidance on corrosion is given in Clause 2.6 and Annex F of BS EN 1993-5 and the 8th edition of the *Piling Handbook*. Protective measures such as coatings, cathodic protection and increased steel grade or thickness to increase design life, are also described.

C12.6 Welding procedures

Figure C12.1 provides guidance on the inspection and welding regime to be adopted with recommendations for testing and acceptance. It is

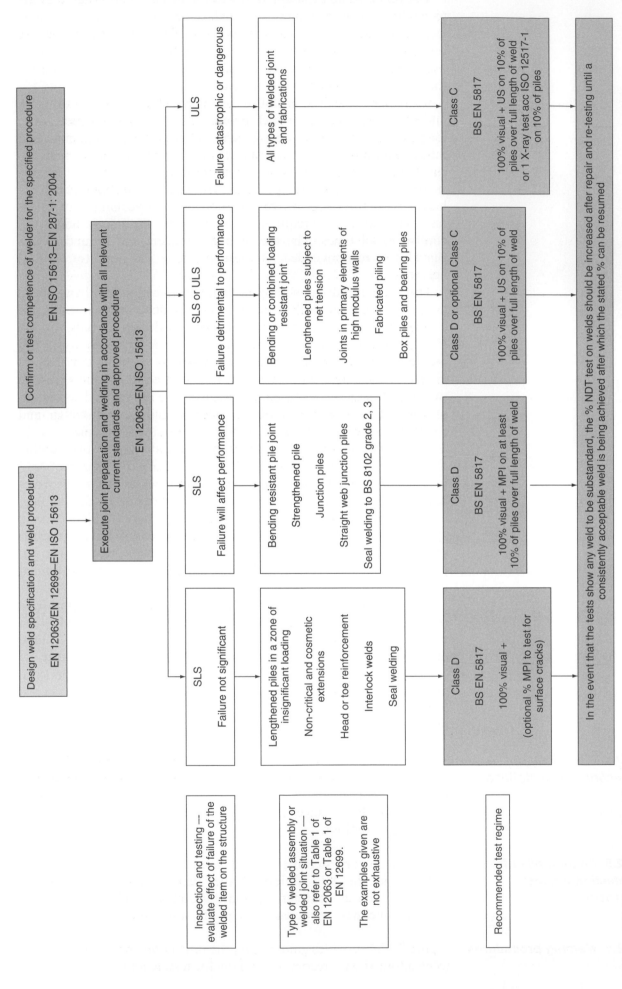

Figure C12.1 Inspection and welding regime

Specification for piling and embedded retaining walls. Thomas Telford, London, 2007

important that the Designer relates the acceptance criteria to the significance of failure of the welded sections.

The Arcelor publication *Welding of Steel Sheet Piles* offers recommendations and guidelines on welding problems encountered when making the more common assemblies.

C12.6.1 Welding standards

Information regarding welding is given in Section 8.4 and Annex B of BS EN 12063. Welder approval to BS EN 287-1 is necessary to demonstrate competence in general welding techniques for both butt and fillet welds. If a specific range of approval is required, it should be clearly stated in the Project Specification.

The welding standard is also appropriate to any associated steelwork such as bracing and strutting to the sheet piling.

C12.6.2 Seal welds

Seal welds can be specified to provide a method of sealing the interlocks of sheet piles and further guidance is given in the 8th edition of the Arcelor *Piling Handbook*. Unless specified with a given throat thickness, seal welds will be made with the minimum number of passes required to bridge the gap between the parts to be welded. There will not normally be any problem with the life expectancy of the structure as the structural form means that the corrosion rate will not normally be high and, in general, the steel from which the sheet piles are made will tend to corrode in preference to the weld. When interlocks have been seal welded, a non-destructive test can be carried out to check the weld integrity and prove watertightness of the joint before water pressures build up to design levels.

C13 Integrity testing

C13.1 General

The purpose of integrity testing is to identify anomalies in piles, barrettes and walls (test elements) that could have a structural significance with regard to the performance and durability of the test element. Integrity tests do not give direct information about the performance under structural loads.

C13.2 Methods of testing

Integrity testing of pile and wall elements is extensively discussed in the technical literature, e.g. CIRIA Report R144.

It must be appreciated that the methods referred to in Section B13 measure acoustic properties of the concrete and from these infer the condition of the test element. Even in simple cases the wave patterns produced are complex. Individual features of test element and variable ground conditions cause further complication and create uncertainty in the interpretation. The results should be viewed in the light of the construction records and knowledge of the ground conditions before conclusions are drawn. The interpretation of tests on elements of completely unknown length requires particular caution since the results can be misleading. The following methods are described in the Specifications:

(a) Impulse response method: The impulse response is a stress wave reflection method for individual piles or barrettes which relies on the measurement of both stress wave reflections and low-strain impact force induced by an impact device (hand held hammer or other similar type) applied axially to the pile normally at the pile head.

(b) Sonic echo, frequency response or transient dynamic steady-state vibration method: These test methods are procedures for determining the integrity of individual piles or barrettes by measuring and analysing the stress wave velocity response and acoustic properties of the pile induced by an impact device (hand-held hammer or other similar type) applied axially to the pile, normally at the pile head.

Impulse and sonic echo methods of testing are not likely to give meaningful results for walls because the signal will be affected by adjacent wall elements. Even with contiguous piles, concrete overbreak is likely to result in some piles touching which again will affect the signal and give misleading results, therefore these methods are not recommended for walls.

There is normally a limit to the length/diameter ratio of a pile/barrette which can be successfully and fully investigated using the above methods depending on ground conditions.

(c) Cross-hole sonic logging: The test measures the propagation time and relative energy of an ultrasonic pulse between parallel access ducts installed in the pile or wall. Both the time between pulse generation and signal reception ('first arrival time' or 'FAT') and the strength of the received signal give a relative measure of the quality of concrete between transmitter and receiver.

Cross-hole sonic logging can be used not only for piles and barrettes but also for large wall elements to determine anomalies in the concrete section. The logging tubes may be formed from either steel or PVC. However, the PVC tubes could develop a debonding defect, particularly in recently cast elements. Since the tubes are fixed to the reinforcement cage, the system will not reveal any information about the integrity of the concrete cover outside the reinforcement cage.

Characteristics of the toe of the wall element cannot be reliably detected with this method unless special measures such as investigation by coring through access tubes below the toe of the wall are adopted. This is not recommended unless a defect is suspected because of the potential risk of disturbance of the founding material.

It should be noted that not all acoustic anomalies revealed by cross-hole sonic logging represent a structural defect in the wall element. The rating of the shaft integrity considers the increases in 'first arrival time' (FAT) and the energy reduction relative to the arrival time or energy in a nearby zone of good concrete. Such anomalies must be judged in the light of all the available data (ground conditions, construction records, etc.).

Acoustic anomalies can arise from:

- insufficient bond between the tube and concrete
- misalignment of the tubes
- lack of fit of the probes in the tubes
- bleeding of concrete.

C13.4 Age of test elements at time of testing

Generally, depending on concrete admixtures, test elements should be at least five days old before testing. Premature testing can lead to erroneous results. A shorter period can be specified with a provision that the time may have to be extended if the first test results are inconclusive and tests repeated. It is best practice to carry out a large proportion, if not all, of the testing while a piling/diaphragm walling rig is still present on site. Should any perceived problems arise, remedial actions, if considered necessary, can be more quickly dealt with.

C13.5 Preparation of element to be tested

Integrity tests are quick — a large number of piles or barrettes can be tested very rapidly (10–20 per hour with good access) — however, the quality of recorded data is directly related to the preparation of the concrete surface and access to the test element.

C13.8 Report

The report will contain all recorded test data, the interpretation of the test data and the inferred condition of the concrete. The inferred condition will result from the analysis of anomalies identified in the test results. It is important to remember that interpretation of these anomalies requires a high degree of judgement and/or subjective interpretation. It is important that all parties recognize the degree of uncertainty of the interpretation.

All recorded test data will include all basic captured data which may be in the form of, for example, millivolts output from transducers, and should include details of any data logging and signal conditioning used, and any field notes.

C14 Dynamic and rapid load testing of piles
C14.1 General

Dynamic load (DL) testing and rapid load (RL) testing are much quicker to complete than static pile testing since they do not require any pre-installed reaction systems. They can be considered as either alternative or complementary methods to static load testing in assessing pile performance provided their results and the interpretation of the results are fully understood. DL and RL testing may be used for site specific quality and consistency work where a maintained (see Section B15) load test is available on an identical pile. DL and RL testing are now well documented and the Engineer should be satisfied that the chosen pile test regime meets the project requirements.

DL tests involve striking a pile with a hammer (for driven piles it is usually the same one as is used to install the pile) or equivalent and observing the resulting forces and motions recorded by gauges fixed near the pile head. For conventional hammer impact tests the duration of motion is of the order of 50 ms or less. The DL test can be completed to ASTM D4945 Standard Test Method for High-Strain Dynamic Testing of Piles or other non-UK national standards.

RL testing is carried out by applying a dynamic load at the pile top through the use of a fast burning material in a confined cylinder and piston arrangement and a reaction weight. The weight is accelerated at around 20 g and the resultant reaction force is the load applied to the pile. The duration of loading is several times longer than the DL tests in the order of 100 ms to 200 ms. Pile head load is measured using a load cell and pile head vertical movement measured using a laser or similar system and/or deduced from accelerometers fixed to the pile. The load and deflection measured during the test are plotted to give pile head load versus deflection. There are no accepted European standards as yet for RL testing.

The main issues that arise for DL and RL tests are as follows:

(a) The resistance measured is 'dynamic' and not 'static'. The link between static and dynamic resistance depends on pile velocity, pile inertia, soil inertia and soil type and properties. Additional contributory factors include the pile dimensions, hammer weight or reaction mass, hammer drop, anvil or cushion condition, propellant mass, age at test, etc. The question of correlation may not therefore be answered by a single formulation.

(b) Rate effects in general enhance apparent pile capacity. This is readily observable, for example, in the case of constant rate of penetration tests. There are many empirical correlations for the velocities that develop when piles are subjected to a dynamic or rapid load. However, currently these correlations are not extended to enough pile types and ground conditions. Hence, dynamic and rapid load tests for piles, particularly in saturated, fine grained or laminated soils, or ground with hard bands, are to be treated with caution, and calibration with site-specific static pile load tests is desirable.

(c) The test duration precludes any measurement or assessment of settlements due to consolidation or creep. These factors can be of significance in normal static maintained behaviour dependant upon ground conditions.

(d) Whether the full dynamic resistance of a pile can be reached depends on the pile movement achieved. It is generally recommended that driven piles be tested some time after normal driving and after some 'set-up' has occurred, hence it may not always be possible to move the pile adequately with the same hammer, even when using a greater drop height. It should be recognized that, even with increased drop heights, the full or 'ultimate' dynamic resistance may not be achieved unless a larger hammer is used.

(e) Piles may be of substantial length. Elastic shortening plays an important part in pile head displacements under static loading but under normal dynamic hammer impact, elastic shortening is a function of the length and form of the compression wave, which might for example be several tens of metres long and with variable amplitude along its length. Elastic shortening therefore has to be inferred rather than measured and because of the pile length and stress level it can in practice become a very important component of behaviour. Many of the published inferred agreements between static maintained load and dynamic tests are found to owe more to the calculation of elastic shortening than any other factor. These features mean that dynamic load testing is mainly a comparative exercise and should not be considered as equivalent to the direct static maintained loading of piles. For rapid load tests the effects of elastic shortening are not well researched at present.

It is advisable, where possible, to use specialists that have UKAS, BSI or equivalent accreditation. This will ensure that the calibration of the sensors and test procedures will comply with a previously agreed standard.

C14.3/C14.4 Construction/ preparation of a pile to be tested

Piles to be tested need to be designed and constructed not only to take the geotechnical loads of the structure but also the dynamic loads from the testing. During DL and RL tests the pile is subjected to forces that may exceed those the pile is designed for, particularly in saturated soils or clays. The provision of additional reinforcement in the pile should be considered to prevent damage to the pile during testing. This is particularly important where piles are being tested for potential re-use.

Note that since rotary bored and CFA piles are often not reinforced for their full length, DL or RL testing may be inappropriate. If testing of rotary bored or CFA piles is considered necessary then test piles must be suitably reinforced. This may make dynamic and rapid load testing unsuitable for random testing on working rotary bored and CFA piles.

In the case of cast-in-situ pile tests, the shape of the pile may have significant influence on the test results.

Preparation of the pile head
Steel and pre-cast concrete piles will not usually require significant preparation. This clause refers more specifically to cast-in-situ concrete piles that will in general require the pile head to be designed to take the higher dynamic loads from the DL or RL tests.

C14.6 Measuring instruments

DL and RL tests are very quick events and the measuring instruments need to be able to respond appropriately. The settlement of the pile is usually calculated by the double integration of data from accelerometers attached to the pile head. This is not a foolproof method of determining the settlement and it is advisable to have an independent displacement measurement system in place. As a minimum it is recommended to measure the level of the pile head before the test, after each cycle and when the test equipment has been removed. The simplest way is to use a precise level method (see Section B17). Other systems that require equipment to be attached to the pile will need to withstand the effects of dynamic loading during the testing and be referred to a datum outside the influence of the tested pile and testing equipment operation.

C14.7 Hammer or propellant

To prevent pile damage during a dynamic load test, the current practice is to limit the resistance mobilized in the pile to the order of twice working load. Where higher resistance forces are considered necessary, a wave equation analysis, carried out prior to pile construction, is very useful as an indicator of the magnitude of dynamic impact force required, and of the potential stresses that may be induced by the impact. Suitable hammer weights, drops, cushions and pile reinforcement can then be detailed.

During rapid load testing the pile is in compression and there are no tensile stresses induced. Hence, during a rapid load test a mobilized resistance higher than twice working load may be acceptable. For rapid load tests the pile head stresses should be assessed in a similar manner as that for static pile load tests.

Often it is appropriate to carry out both static and dynamic or rapid load tests on the test pile and to compare the load–settlement curves obtained for the pile from each test type. In such a case care is required over the timing and sequencing of comparative testing on the same pile as this may have an influence on the results achieved.

C14.8 Time of testing

Generally, the minimum time between the completion of installation and testing for any pile shall be at least four days to allow for excess pore water pressure dissipation and, for cast-in-place piles, the development of concrete strength.

However, there are two cases where there is no need to allow for excess pore water pressure dissipation or for the development of concrete strength:

(a) Testing of preformed piles may also be used to investigate hammer performance and/or pile stresses. Under these circumstances it is normal practice to carry out testing during the driving operation. Dynamic testing can be used to observe changes in resistance forces with time, and testing at time of installation can be of use to the Engineer.

(b) Testing of preformed piles that found in coarse grained soils or into rock may need only 12 hours for any excess pore water pressure dissipation.

C14.9 Interpretation of results

The interpretation of DL and RL tests is continuing to be developed and should be carried out by specialists who are up to date with the latest information. The interpretation of both types of test have seen rapid development and research is continuing for application to all situations and soil types.

The interpretation of a DL test should include a one-dimensional wave mechanics theory that allows the data to be analysed instantaneously for estimate of pile test load; other important information immediately available is pile integrity, pile compression and tension stresses, and hammer blow energy. Further analysis of the test data can be carried out using stress wave theory and wave equation based signal matching analysis programs. The current empirical relationships will only work in soil/pile types for which they were developed and care should be taken when applying them to other situations.

The interpretation of RL tests is usually completed by the specialist carrying out the test using procedures developed by the manufacturer of the equipment or other techniques published in the literature. The interpretation is less straightforward because the rate of loading of the pile varies during the test. Therefore, if the soil in which the pile is founded exhibits a loading rate effect, the comparison of the RL results with static tests will be difficult to achieve. The current empirical relationships will only work in soil/pile types for which they were developed and care should be taken when applying them to other situations.

C14.10.1

(*h*) Other relevant details may include rig type, dead load imposed at start of test, equipment details, etc.

C14.10.2

(*j*) All recorded test data will include all basic captured data which may be in the form of, for example, millivolts output from transducers, and should include details of any data logging and signal conditioning used, and any field notes.

C14.10.2 Additional information

For preformed driven piles it is good practice to always measure the set at testing since the set value may well be taken into consideration in the dynamic analysis. Where pile set at testing is compared with the pile set obtained at installation, any differences that may exist in hammer performance or operation should be taken into consideration. It may be the case that the set at testing appears more open and this may simply be due to a greater impact force at testing resulting in a more open set.

C14.10.3 Analysis

There are several commercially available software packages to analyse DL and RL pile tests. It is important that the software package applied to the data is identified and the parameters used in the analysis are given, together with the source of the values. The source of the values will include whether they were measured specifically in relation to the soil and the piles tested or are estimated values. Estimated values should be justified and the source quoted.

C15 Static load testing of piles
C15.1 General

This section deals with the testing of a pile by the controlled application of an axial load. It covers vertical or raking piles tested in compression (i.e. subjected to loads or forces in the direction such as would cause the piles to penetrate further into the ground) and vertical or raking piles tested in tension (i.e. subjected to forces in a direction such as would cause the piles to be extracted from the ground). Static load testing of piles is one reliable way to establish a pile's load-settlement behaviour.

For further information on static, lateral and tensile load test methods and arrangements refer to CIRIA Report PG7 and *Handbook on Pile Load Testing* by the Federation of Piling Specialists.

Static pile testing using bi-directional methods are not included in this section. This is a proprietary method economic for test loads in excess of 10 MN.

BS EN 22477 has not been published at the time of publication of this document. It is anticipated that this European Standard will have requirements that are inconsistent with this Specification and UK pile testing practice. This Specification is a document designed for UK practice and its requirements are to take precedence. However, it is expected that specifiers post publication of BS EN 22477 will be aware of its requirements and that differences between BS EN 22477 and the Project Specification will be a deliberate choice.

Preliminary piles

The purpose of preliminary piles is to validate the geotechnical design of the pile, achievable performance criteria and to prove that a Contractor's method of construction can construct viable foundations in particular ground conditions. Where preliminary pile testing is considered necessary and the ground conditions show significant variation from one area to another, tests may be required in more than one location. Indeed, more than one test per location may be necessary to judge pile performance under compressive, tensile or lateral loads.

Where preliminary piles are required, they should be constructed sufficiently in advance of the installation of the working piles to allow time for the test, the evaluation of the results and the adoption of modifications if these prove necessary. If it is necessary to specify a precise timing for the construction and testing of preliminary piles, this should be included in the Project Specification.

Measurement of test piles

Preliminary piles installed and tested before the start of the main works should be measured in a separate bill of quantities or be the subject of a separate contract. Their installation should be measured in accordance with the appropriate method of measurement for the type of pile, and separate items should be measured for each specified loading and type of test. For tests on working piles, separate items should be provided in the bill of quantities for each test. The item description should include all preparatory work such as site surfacing, bringing the pile head to the commencing surface level, trimming the head of test pile and all work required in connection with the form of a reaction system including, where applicable, the supply of kentledge.

Where the Contractor has designed the piles and performance criteria are specified, a test is an essential part of the contract to establish that the piles meet the specified performance criteria.

Table C15.1 Typical pile testing strategy based on risk levels

Characteristics of the piling works	Risk level	Pile testing strategy
Complex or unknown ground conditions No previous pile test data New piling technique or very limited relevant experience	High	Both preliminary and working pile tests essential 1 preliminary pile test per 250 piles 1 working pile test per 100 piles
Consistent ground conditions No previous pile test data Limited experience of piling in similar ground	Medium	Pile tests essential Either preliminary and/or working pile tests can be used 1 preliminary pile test per 500 piles 1 working pile test per 100 piles
Previous pile test data is available Extensive experience of piling in similar ground	Low	Pile tests not essential If using pile tests either preliminary and/or working tests can be used 1 preliminary pile test per 500 piles 1 working pile test per 100 piles

It is desirable that preliminary piles are specified unless the design, factor of safety against failure, construction method and ground conditions are such that the risk of failure of working piles is minimal. Table C15.1 gives an example of appropriate levels of pile testing based on risk.

C15.2 Project Specification

C15.2 (b): This Specification is for axial maintained load tests, but other tests exist which in particular circumstances may be more appropriate such as constant rate of penetration (CRP) or bi-directional pile testing. These guidance notes contain information on these other two types of test.

The type of axial maintained load test should be stated in the Project Specification as:

- proof load test as Clause B15.13.1
- extended proof load test as Clause B15.13.2
- single-cycle or multi-cyclic load test as Clause B15.13.1.

Where other systems of test are stated, details should be provided of all the loading stages, measurement requirements and acceptance criteria.

CRP testing

The CRP test is no longer considered appropriate for the majority of testing situations. This is because the results may lead to apparently enhanced higher capacities by comparison to those derived from maintained load tests due to rate effects. The high rate of loading which lead to those apparently enhanced capacities are not representative of many real situations so its usage should be confined only to situations where dynamic loading effects may be important.

However, in certain testing situations, for example when dealing with a strain softening response, where the pile capacity reduces with displacement, the CRP test is still preferable. It is difficult to determine these characteristics using static maintained load techniques.

If a CRP test is to be used, the following could form the basis of the specification:

Constant rate of penetration (CRP) testing procedure

The rate of movement of the pile head shall be maintained constant in so far as is practicable and shall be approximately 0.01 mm/s for piles in predominantly cohesive soils and 0.02 mm/s for piles in predominantly coarse grained soils.

Readings of loads, penetration and time shall be made simultaneously at regular intervals; the interval chosen shall be such that a curve of load versus penetration can be plotted without ambiguity.

Loading shall be continued until one of the following results is obtained:

(a) the maximum required test load as specified in the Project Specification is reached

(b) a constant or reducing load has been recorded for an interval of penetration of 10 mm

(c) a total movement of the pile head equal to 15% of the base diameter, or any other greater value of movement specified in the Project Specification has been reached. The load shall then be reduced in five approximately equal stages to zero load, with penetration and load at each stage and at zero load being recorded.

Bi-directional method

Another form of maintained load test is the bi-directional method. This system is normally applicable to rotary bored piles. A preformed load cell is placed in the pile bore either at the pile base prior to concreting or part of the way up the pile shaft during the concreting operation. In the test, the cell is hydraulically expanded so the upper portion of the pile reacts against the lower section. Where the cell is at the base of the pile, the soil at the base provides the reaction. More than one cell can be provided to test different sections of the pile shaft.

There must, however, always be sufficient shaft resistance from the pile section above the load cell to provide the necessary reaction force to stop the pile being forced out of the ground. The load is quantified by measurement of the hydraulic pressure of the jack cast into the pile.

Interpretation of the results is needed to derive the equivalent load/settlement characteristic and this can only be undertaken by persons experienced in this type of test. The test arrangement needs to be carefully designed to allow meaningful interpretation.

C15.2 (h): Instrumentation can be specified either as devices at particular locations or levels; or as a general requirement to establish for instance elastic shortening over a particular length. For further information see Section B17.

C15.2 (q): To test the lateral capacity of a pile, the typical method would be to jack apart, or pull together, two adjacent piles. This method obviates the need for the provision of any additional reaction system, however, both adequate structural performance and an acceptable deflection performance need to be ensured. Alternatively a separate reaction system could be designed.

Often test arrangements do not adequately model the real loading condition and the results therefore are of little value. This is particularly the case where additional lateral restraints such as pile caps, floor slabs, bridge piers, pile groups or basements will be present in the permanent works.

C15.3.2 Method of construction

Special construction details may be required for preliminary piles in order to provide data that are relevant to working piles, say where downdrag is expected or where piles will form part of a deep basement

---

or substructure. Details may include sleeving and instrumentation within the pile.

The method of construction for preliminary and working test piles should otherwise replicate as closely as possible the methods used to form all the working piles.

As preliminary piles are test loaded to much higher loads than the working piles will ever experience it is often necessary to increase the structural strength of the preliminary pile so that it can safely withstand these forces.

C15.3.3 Boring or driving record

Caution should be taken that special testing during construction, for instance taking samples or doing in-situ tests, may make the pile behave differently to working piles and hence affect subsequent behaviour.

C15.4 Concrete test cubes

Where test piles require significant quantities of concrete, additional sets of four test cubes should be taken.

A calculation should be undertaken to calculate the design concrete strength needed for the test pile at the time of testing, and this should be related back to a concrete cube strength. A rule of thumb is that the concrete cube strength should be twice the average stress applied to the pile and the cap at the maximum test load.

C15.5 Preparation of a working pile to be tested

The cap should be cast as an integral part of the pile whenever possible.

C15.6 Cut-off level

It is important that the Design Verification Load is appropriate to the situation of the test and the long term loads for which the piles are being designed. As an example, if downdrag is expected on the working piles, twice the expected downdrag force should be added to the Specified Working Load to give the Design Verification Load, once to overcome the positive skin friction over the relevant length and a second to replicate the actual downdrag loading. Where working piles are being installed in advance of an excavation, the Design Verification Load should take account of the support provided by the soil that will be excavated and also by the higher effective stresses giving higher strengths in the soils beneath.

C15.8 Safety precautions
C15.8.4 Testing equipment

Other than carrying out the load test it is not possible to check on site that the hydraulic test equipment is capable of sustaining the required pressures without leaking. It should also be noted that in practice the likelihood of leakage occurring from the jack is a function of the ram displacement. All fittings and pipes should normally be rated for 1.5 times the maximum required pressures and the equipment should be capable of safely sustaining the maximum test load.

C15.9 Reaction systems
C15.9.2 Compression tests

The weight of kentledge blocks is often over-estimated leading to under-capacity. Contractors who choose to supply a weight of kentledge only 10% greater than the maximum test load will need to be able to demonstrate without doubt that the test will be safe.

For any kentledge that becomes potentially unsafe, the Contractor will need to take measures to resolve any issues with bearing capacity and will need to consider stability of the stack.

Guidance on pile load testing procedures is given in CIRIA Report PG7. The methods of providing the reaction, i.e. kentledge, reaction piles or ground anchors, should be submitted by the Contractor and not specified by the Engineer except in some instances for driven piles where the uplift of the test pile during driving of the reaction piles is of interest. Safety of the test arrangement is of paramount importance and the Contractor should give the Engineer the specified period of notice so that the test assembly may be inspected by the Engineer before the application of any load.

C15.9.4 Working piles

Although the Contractor may propose the use of working piles as reaction piles, the Engineer should state whether or not this will be acceptable in terms of their incorporation and performance as part of the permanent works. Where working piles are used as reaction piles, the fundamental checks of working piles are:

- that the strains in the pile at maximum load do not impair durability
- that the subsequent load–settlement performance is not significantly adversely affected.

C15.9.5 Spacing

The use of kentledge may influence settlements in the surrounding ground. This may affect the reference beams as well as the test pile. An independent checking system (e.g. precise levelling) will highlight gross changes in reference beam level.

During a compression test, the reaction piles may unload the ground surrounding the test pile and potentially yield a lower capacity. The opposite may occur in a tension test whereby the reaction piles increase the effective stresses in the ground around the pile and may lead to a higher test capacity being measured than normal working piles. It is for these reasons that reaction pile spacings are kept bigger than economically desirable, and especially so for tension tests.

There is no clear evidence that deep reaction piles constructed at three pile diameters from the test pile have any significant adverse effect on the test result. If the Contractor proposes a spacing of less than three diameters, then it should be demonstrated that there will be no significant interaction between the test pile and the reaction piles.

C15.10 Application of load
C15.10.1 Equipment for applying load

The Engineer should satisfy himself that the Contractor's method of load application will only impart an axial load into the pile, unless the load is on a laterally loaded pile.

Shims may be used to extend the travel of a short-stroke jack if safe to do so.

C15.10.2 Measurement of load

Whilst manually controlled tests are acceptable for the most basic preliminary and proof load tests, more detailed analyses of pile test results requires closer control of the test load. This is only likely to be achieved with some measure of automation. This allows for load to be controlled with greater consistency and for readings to be taken at closer intervals than specified if deemed necessary.

Fluctuations in load during hold periods can distort the recorded pile behaviour, resulting in variations in the apparent rates of settlement.

The load cell's routine calibration programme would normally be part of a Quality Assurance system, with calibration intervals normally of 12 months. Significant drift tends to be more unusual than changes due to instrument damage. If the measured load–settlement performance appears unusual, a recalibration may be appropriate. Devices should always be recalibrated before use when repairs or damage have occurred. The check on hydraulic jack pressure is an approximate verification that the correct calibration has been applied.

C15.11 Measuring pile head movement

Other methods may either be required or proposed as an alternative. The key requirement is for independent and reliable systems of sufficient accuracy and precision for later interpretation.

C15.11.2 Reference beams and displacement transducers/dial gauges

The requirement for the reference beams to be supported at least three pile diameters from the centre of the test pile, or 2 m from the edge of the test pile, is reasonable for most tests. However, reference beams of greater than say 5 m span tend to display significant thermal or vibration influences on the measurements. When testing larger diameter piles the reference beam length and testing procedures may need to be amended or the reference system instrumented so that thermal and vibration influences can be calibrated out.

C15.12 Protection of testing equipment

The temperature can vary considerably from shaded points to areas in the sun. Temperature should therefore be recorded at a point close to where measurements are actually being taken. It is advisable to shade the test area from sunlight in order to minimize temperature fluctuations.

Vibrations may disrupt the test. At the beginning of the test, it should be checked that the measuring system is relatively unaffected by on-site vibrations. It may be necessary to establish an exclusion zone and on confined sites work may have to cease during pile tests as the influence of vibration on the results may be detrimental. Any unexplained fluctuations during the test should be investigated.

C15.13 Test procedure for maintained load compression test
C15.13.2 Proof load test procedure for working compression piles

Working piles should not be deliberately tested to failure. Sometimes failures do occur at loads lower than anticipated and the Contractor and Engineer will need to react accordingly. Generally, working piles are tested to verify that the construction methods used have not changed so as to produce piles inferior to the preliminary piles, and/or to verify that piles, which are for some reason suspect, have acceptable load–settlement characteristics.

A working pile may be tested at any time during the contract but the potential benefit of choosing a pile from several that have been completed must be balanced against the risk of delaying the remainder of the Works if the pile proves to be unsatisfactory. The method and timing of test pile selection should be agreed between the Engineer and the Contractor prior to the Works. For all cases, where pile testing is specified, it is preferable to test at least one working pile in the early stages of the Works to be confident that the performance requirements of the Specification have been met.

However, where cost and time constraints make testing inappropriate, and where the ground conditions are well understood, the factor of safety can be increased to reflect the reduced certainty in pile performance.

There are advantages and disadvantages regarding whether test piles are pre-selected or not. Overall, it is desirable not to pre-select piles so that they are truly representative, but provided the Contractor builds the pre-selected test pile in the same manner as the other piles and this can be verified independently, then there should be no fundamental objection to pre-selection. The potential reasons for each choice are summarized in Table C15.2.

It will be noted that the minimum load holding period recommended in Table B15.1 has a six hour period at 100% DVL and at

Table C15.2 Reasons for test pile pre-selection

Reasons for pre-selection of piles for testing	Reasons for post-selection of piles for testing
1. When special reinforcement details needed	1. Avoids suspicion that test pile has been built to be better than typical
2. When instrumentation is required	
3. When temporary anchor pile layout restricts choice of location	2. Reaction to particular events during construction of one pile or for a family of piles, or the discovery of different ground conditions
4. When site layout restricts choice of location — for instance so that there will less programme pressure to complete the test quickly	
5. Allows test pile cap to be cast at same time as pile, for improved strength for testing to high loads	
6. Test pile may be amongst first piles installed, which reduces consequences if poor results are received after most piles have already been installed	

100% DVL + 50% SWL, whereas other load holding periods are potentially significantly shorter.

The effect of this is to cause consolidation and creep to be concentrated into the results of some increments while displacements at other loads may be significantly smaller. The result is that if interpretation is done by simply joining the final displacement points at each load, a somewhat erratic line is produced. In particular, the steepness of the line obtained by joining the final settlement points for the last two stages might lead one to anticipate an imminent failure or ultimate condition when this may not be the case.

The long term final load–settlement curve demonstrates that a different and correct interpretation is available provided the data has been acquired with accuracy and interpreted according to the best available techniques.

Cycling of test load is now optional as it is not representative of loading conditions for most structures. Where cycling of load may be relevant for the design of a structure or where pile recovery from Specifed Working Load is required, then consideration should be given to optional cycling of load during the test as provided in Table B15.1.

It is acceptable for a time break to occur at the end of any load cycle. For example, the test could be loaded up and down for cycle 1 in day 1, left unloaded overnight, and then cycle 2 carried out on day 2, etc. In future, this may make pile testing easier to comply with the Working Time Directive.

In the event that the primary settlement measuring system develops a fault and the secondary settlement monitoring system is being relied upon to complete the test, the periods of time for which loads must be held constant to achieve the specified rates of settlement shall be extended as necessary to take into account the lower levels of accuracy and to allow correct assessment of the settlement rate.

C15.13.3 Load test procedure for preliminary compression piles

Minimum load holding periods

Minimum load holding periods are specified in Table B15.1, but these are modified by Clause B15.13.1 where limiting rates of settlement have also to be satisfied. The settlement rates recognize that as loads become higher and settlements become larger, a pile under load takes longer and longer to reach a near stable state. If the chosen applied load should be the ultimate load, it would take an infinite time for a state of stability to be reached. This Specification attempts to address this issue but the formulation given may not always be the most appropriate.

In practice what is required from an analysis is to establish enough of the time/displacement curve to allow prediction of the final settlement for the load under consideration. This would rarely require load hold periods in excess of six hours, even for quite large diameter end-bearing piles.

Recycling load at working load stage
It is understandable to want to quantify the effect of load variation about the working state when the supported structure is in service. It is nevertheless only in rare cases where load fluctuation is possible between the working and zero load states. Large fluctuations may occur in the case of material storage silos or where piles may undergo alternating compression and tension forces because of their structural function.

C15.15 Presentation of test interpretation

Contractor interpretation of the pile test results is generally only relevant where the piling subcontractor has design responsibility for the piles.

If interpretation of pile test results is specified in the Project Specification along with Option 1 'Engineer design', the aims of the test and the pile design basis must be carefully explained by the Engineer in order that the Contractor is able to carry out a meaningful interpretation of the test.

Specification for piling and embedded retaining walls. Thomas Telford, London, 2007

C16 Piles with sleeves and/or with coatings
C16.1 General

Sleeves may also be referred to as casings or liners.

The reduction of friction using sleeves and/or coatings may not be the most economic or best practical solution. The Designer should consider, at an early stage, whether a more efficient solution could be adopted (e.g. increased pile capacity to cope with negative skin friction). The Designer must ensure that the chosen technique and materials are compatible with the design, particularly where lateral loading is to be resisted.

Particular methods for reducing friction or for providing a protective barrier usually comprise coatings for driven piles or permanent sleeves, either with or without coatings, for bored piles. A variety of coatings may be used including bituminous material, sintered metal paints and thixotropic suspensions such as bentonite. Systems of concentric casings incorporating grease layers are also available.

There are often practical difficulties in installing such coatings or sleeves and so, where possible, it is preferable to avoid their use and design the piles accordingly. This approach reduces the risk of poor workmanship and can reduce costs.

Sleeves and coatings may be required for a variety of reasons. They are commonly employed to reduce friction and thus minimize the effects of clay heave or negative skin friction. However, they are also used for other purposes (e.g. protection against chemical attack in aggressive soils or to deal with voids in the ground). Hence the Engineer should make clear not only the full details of the protection system but also its purpose in order for the Contractor to assess the practicality of installation and/or offer suitable alternatives. The Engineer should ensure that the system used is compatible with the design.

C16.3 Bituminous or other coating materials
C16.3.2 Protection from damage

Coatings may need oversized leading edge protection to minimize potential installation damage.

Care should be taken when driving piles with pre-applied coatings through coarse granular soils that the coating is not damaged or removed.

C16.5 Inspection of coatings

If the Engineer specifically requires piles to be partially exposed or extracted then this should be made clear within the Project Specification as it is can prove extremely difficult to extract piles as special equipment is normally required. This, coupled with the risk of being unable to complete extraction, make this operation difficult for the Contractor.

During installation, some damage to coatings applied to piles or sleeves is inevitable. This fact must be addressed by the Designer who should consider the likely damage that the installation process may inflict on the coating or sleeving. Discussion between the Designer and the Contractor on this matter, prior to work on site, should avoid any later misunderstandings. In this respect, trial piles may be beneficial to both parties to ensure that the proposed piling system and the Engineer's design requirements are satisfactorily met.

C17 Instrumentation for piles and embedded retaining walls
C17.1 General

This section specifically applies to instrumentation for piles and for embedded retaining walls. The instrumentation can be used for monitoring stresses or displacements either in preliminary load tests or in service and for lateral and vertical loads. The results can be used to derive the load and deflection distribution down the pile or wall.

C17.2 Project Specification

The Project Specification must include details of the aims and objectives of the instrumentation and appreciate the impact that the data can have on the project. This will be reflected in the range and resolution of the instrumentation selected.

Thought should be given when specifying instrumentation as to what needs to be measured, whether those measurements give the required behaviour and how exactly the results will be used. As an example, if inclinometers in a retaining wall are being monitored, it will be necessary to know also the depth and extent of the excavation at each set of readings.

If the instrumentation is to be used as part of the 'Observational Method' then the readings, analysis and interpretation of the data will be on the critical path of the project and must be completed in the specified lapse time in order to trigger the next action in the construction process.

It is important to coordinate the schedule of monitoring instrumentation with the schedule of construction activities. Where unexpected and anomalous readings are obtained, the schedule of construction activities should be checked against the actual construction activity at that time whilst the instrumentation monitoring is repeated.

The location of instrumentation and monitoring stations should be considered in the light of the whole construction process, particularly if there are temporary works, to ensure that the instrumentation can work as required.

Consideration should be given to the inclusion of redundant instrumentation to account for instrument malfunction and as back-ups and checks.

It is essential to assign the responsibility for instrumentation to ensure it is read, maintained and the data analysed. It is very important that one person is responsible for all aspects of instrumentation and has the time to coordinate all the necessary activities and interface with other works.

C17.3 Type of Instrumentation

Careful consideration should be given to the selection of the type of instrumentation. Many systems available have proven track records in piling and retaining wall works while newer systems may not. It is important to get information on systems from more than one source, i.e. manufacturer and user, installer and user. Some types of instrumentation are readily read by remote reading systems that transmit the data to a collection point that can be in the office of the responsible person or the Engineer. However, other types of instrumentation can only be read manually so the selection of the type of instrumentation needs to consider the reading method.

C17.3.1 Extensometers

Extensometers generally measure the changes in length of a pile or sections of a pile. There are two main types of extensometers. One is a permanently installed rod type where the rod is protected from influences of the pile, with one end connected to the pile and the other to a displacement measuring device. The other type uses an access tube, cast in or fixed along the length of the pile, for a sensor to travel along and locate targets previously fixed in the pile. Both types need access to the top of the pile for setting up the reading systems and while automatic reading systems are available for the target seeking type they are not as compact and popular as those for the rod types.

C17.3.2 Inclinometers

Inclinometers measure the lateral displacements relative to the head or toe of the pile or retaining wall. This can be achieved either by a torpedo that senses the inclination of each successive length as it is lowered or raised in the access tube or by a system permanently installed in the access tube, such as the electro-level string. As with the extensometers the choice of system will be decided by speed of reading, level of automation and access to the top of the pile or retaining wall. The torpedo system requires access at the specified reading times to the top of the access tube and manual lowering and raising of the torpedo, stopping at each reading level to take the reading. The electro-level or equivalent string system would be permanently installed and could be read remotely and relatively rapidly both at predetermined times and other times should the need arise.

It is important to agree and establish the datum profiles of the access tubes before the instruments can be used to monitor movements caused by construction activities. The monitoring data will be compared to the datum profiles at each stage of construction to determine the horizontal movement of the access tubes.

C17.3.3 Load cells

Load cells are used to measure axial load, either tensile or compressive, and their location and installation must not compromise their action. Load cells are most commonly used in piles for preliminary tests and their calibration over the required load range should be checked for variation with temperature, particularly if they are to be used with an automatic load maintainer type system that requires output to control the loads applied. As load cells are moved from pile test to pile test, this increases the likelihood of damage and so the latest calibration certificate and maintenance record should be available before the test for approval.

The installation of load cells in piles during their construction is extremely difficult. For example, where the total vertical toe load on a bored cast-in-place pile is to be measured, the installation must ensure that the load cell is not bypassed by concrete that will take load, so reducing the measured load.

C17.3.4 Pressure cells

Pressure cells are one of the more difficult cells to install, particularly for concrete–soil interface or existing concrete–new concrete interface. The bedding of the measuring faces to the existing material need careful attention if good quality data is to be obtained. The pressure cells should be calibrated when bedded in the same way as they will be bedded in the piles.

C17.3.5 Strain sensors

Strain sensors cover a range of different systems from the traditional short gauge length (the order of centimetres) sensors with vibrating wire or foil strain gauge based strain measurement to the newer long gauge length (the order of metres) systems based on fibre optic technology. Each system is read with different readout equipment and has different response times and reading times. For example, foil strain gauges have rapid response, fast enough for dynamic readings, whereas vibrating wire systems take longer to read (still only a matter of seconds) but can only be used for static conditions. Fibre optic systems can give a continuous strain profile down the length of a fibre optic cable but will take some minutes to get the data and require expensive readout equipment to do it. Other fibre optic based systems are effectively short gauge length sensors that can be read rapidly. It is vitally important therefore that the aims of the instrumentation are matched to the capabilities of the selected sensors.

Strain sensors should be placed so that they are evenly distributed both with depth and, in the case of an embedded retaining wall, lengthwise along each panel. Serious consideration should be given to including extra redundant sensors to provide both duplicate readings and security, should sensors or their associated connectors and readout units become damaged.

Care should be taken with fixing strain sensors to steel to ensure that the type of fixing system will work for the life of the Project and will not either distort or damage the sensor.

C17.3.6 Surveying

Surveying can be used for measuring vertical differential movement, precise levelling, or, for spatial movement, geodetic surveying. Precise levelling circuits need to be designed to suit the equipment used, preferably with equal fore-sight and back-sight distances and fixed change points. Precise levelling cannot be completed remotely or automatically so is best suited to checking tasks and the calibration of other instrumentation or situations where the datum for measurements is some distance from the works. Where the settlement behaviour at, say, a number of times during a construction project is required then precise levelling can give settlements to within 0.5 mm. If used for checking the settlements during a pile test with a simple backsight on to a datum, then significantly better accuracy can be achieved. Care must be taken with the assessment of capabilities of instruments that may have a high resolution but the levelling circuit will significantly reduce the accuracy of settlements.

Modern equipment for geodetic surveying can be set up for automatic reading, data storage and transmission of data to a remote monitoring station. These systems are relatively slow but can work well where only a line of sight is possible and spatial movement is required. As with precise levelling the resolution of the instrument will probably not equate to the accuracy of the measurements.

C17.3.7 Other Special Instrumentation

This category of instrumentation is for any instrumentation included in the Project Specification that cannot be included in any of the previously mentioned sections, typically newly developed instrumentation. For specialist non-standard instrumentation the knowledge and reputation of the supplier and installer needs to be considered with, preferably, case studies to show the performance of the instrumentation in a similar circumstance to that proposed. The same properties as for all other instrumentation must be examined such as calibration, effects of temperature, durability, long term behaviour, etc.

C17.4.2 Protection

If the Project Specification includes monitoring the instrumentation during construction and use of the building, the location of the instrumentation terminal protection needs to be carefully considered and built into the project planning to be assured of uninterrupted access during the reading schedule.

Equally it is important to notify all the people working in the same area as the instrumentation and the reading point of the instrumentation and make sure it is properly protected. Otherwise, if sufficient care is not taken throughout the installation, reading and interpretation processes, the results can become impossible to interpret.

C17.4.3 Surveying

Surveying in this clause is to show the location of the works in relation to the project grid coordinates and in relation to the project level datum. The data should be reported to provide as built position and levels.

C17.5.1 Monitoring equipment

The monitoring equipment will include all the readout units, data loggers and data transmission equipment used in relation to the instrumentation. For manually read instrumentation it will refer to readout units that should be numbered so the readings can be related to a particular unit. The readout unit should have a calibration chart (if necessary) with appropriate traceable links to a calibration standard held by an accredited laboratory and the date of the last calibration and the calibration interval.

If the monitoring equipment includes an automatic system with transmission of data, site back-up systems need to be in place to ensure no data is lost should a power failure occur. If the unit is battery powered then regular checks need to be incorporated into the reading schedule and battery power should be monitored with the instrumentation data.

C17.5.2 Readings

It is important that all the sensors and monitoring equipment are calibrated and the method and conditions of the calibrations should be recorded with the certificates of calibration. There are many things that can affect the recorded data from a sensor, including temperature differences between readout unit and sensor, connector cable length, connectors, etc. All the variables should be taken into account in the calibration certificate to enable corrections to be made to the readings.

The timing of the readings with the stages of construction can be critical to the success of instrumentation. The type of instrumentation and its reaction time should be considered when designing a system to ensure the required measurements are acquired. Remote reading systems and wireless systems can be affected by the environment close to both the sensors and the remote reading station and these effects should be considered at the design stage. It is essential that all groups working in the same area of the site are aware of the instrumentation and the timing of readings.

C18 Support fluid
C18.1 General

There are three main support fluids used in piling and diaphragm walling and these are water, bentonite and polymer. The selection of the correct support fluid and careful observance of best practice in its use is essential if problems are to be avoided. In the UK there exists a wealth of experience of the use of both water and bentonite over more than forty years.

However, experience in the use of polymer is limited and workmanship issues have arisen on several UK projects. There are a great number of polymers and related additives available and therefore it is essential that the Contractor can demonstrate previous and relevant successful experience under similar conditions. Without such experience, consideration should always be given to a full-scale site trial in order to validate the construction process.

Details of the proposed mix proportions and method of mixing the support fluid should be provided by the Contractor, with details of manufacturer's instructions and certificates, including for any additives. There may be a difference between the manufacturer's recommendations and the Contractor's proposed proportions, especially for polymers. In this instance, the Contractor will need to demonstrate the reasons for this.

Water

In certain conditions water can provide an acceptable support fluid. If water is used as the support fluid, an excess head of at least 2 m above the groundwater table is likely to be necessary to maintain stability. It is therefore important that the supply of water to the excavation is not interrupted.

In silty and sandy clays, and in some mudstones and shales, water can cause swelling and softening of the deposit resulting in loss of stability. In common with some polymers water will not maintain solids in suspension though clays and silts will disperse in the water so that the fluid becomes very viscous. Thorough base cleaning is therefore required before concrete is placed.

Bentonite

There are three common types of bentonite, namely:

- natural sodium bentonite
- natural calcium bentonite
- sodium-activated bentonite.

All bentonites have a capacity to exchange cations, which is much greater than for most other clays.

Most bentonites used in the UK to produce support fluids are sodium-activated, and are predominantly of UK or European origin. Natural sodium bentonite is rarely used because of its high cost. Natural calcium bentonite is not suitable for piling and diaphragm walling.

Typically, if 3% or more bentonite powder is dispersed in water, a viscous slurry is formed which is thick when allowed to stand but thin when energy is applied to it. This phenomenon is known as thixotropy, and results from the orientation of the plate-like particles within the slurry. When the slurry is allowed to stand electrostatic bonding forces between the particles form an interlinked structure which is observable as a gel. When the gel is agitated, the electrostatic bonds are broken and the gel becomes fluid, with the particles orientated at random.

In order to exert stabilizing pressure on permeable walls of an excavation, bentonite slurry must form a seal on the surface of the soil. This avoids loss of slurry into the soil, with consequent increase in pore pressure and reduction in shear strength, and enables the slurry to exert its maximum stabilizing effect.

The seal can be formed by three different mechanisms:

(a) surface filtration
(b) deep filtration
(c) rheological blocking.

Surface filtration occurs when a filter cake is formed by the bridging of hydrated bentonite particles at the entrance to the pores in the soil, with negligible penetration of the bentonite into the soil. During and after its formation, water percolates through the filter cake from the slurry into the soil. Water lost in this way is referred to as fluid loss.

Deep filtration occurs when bentonite slurry penetrates into the soil, slowly clogging the pores and building up a filter cake within them. In this case, the seal may penetrate into the soil approximately 40–50 mm.

Rheological blocking occurs when bentonite slurry flows into the soil until it is restrained by its shear strength. In this case the slurry may flow up to several metres into the soil.

Of these three mechanisms surface filtration is preferred, since the seal is formed very rapidly with no penetration of bentonite slurry into the soil.

Contamination with native soil, usually as a result of ion exchange with calcium or aluminium from the soil, can cause a marked degradation of the fluid and sealing properties of the slurry. Cement contamination has a similar effect and can be detected by an increase in pH. The shear strength of the slurry is important in keeping small particles in suspension and hence avoiding the formation of sediment at the base of the excavation. The replacement of helical reinforcement by hoops or bands of steel spaced well apart is recommended to minimize accumulation of bentonite residue on the intersection of longitudinal and shear reinforcement.

Guidance on these aspects is given in CIRIA Report PG3 which also identifies practical limitations to the construction of excavations under bentonite slurry. Further information on the use of bentonite can be found in the FPS publication entitled *Bentonite Support Fluids in Civil Engineering in the UK*.

Polymers

Polymers can be used as an alternative to bentonite to maintain stability of pile bores or diaphragm wall excavations. Polymers have been used successfully on bearing pile projects in the UK, but experience for diaphragm walling in the UK is limited.

Unlike bentonite slurries, which form a filter cake at the surface of the soil against which the fluid pressure acts, polymeric fluids may form a barrier by rheological blocking of the pores within the soil or by interaction with the soil to form a gel. A feature of some polymers is that they show high viscosity at low shear rates whilst maintaining low viscosity at high shear rates. High viscosity is useful under quiescent conditions to minimize penetration into the soil and to keep cuttings in suspension. Low viscosity is required when the fluid is agitated to minimize pumping pressures, assist rapid flow past digging tools and enhance separation performance during the solids removal stage. The penetration distance into clayey soils is small, but can be significant in sandy or silty soils. Bentonite can be used in conjunction with polymers in granular soils to assist the blocking action.

Polymers can also be designed to inhibit the dispersion of clays and fine soils into the slurry thus promoting lower slurry viscosities and allowing easier cleaning of the slurry as the cuttings remain coarser.

In certain cases (e.g. predominantly cohesive soil, tension piles) it may be preferable to use a low-viscosity polymer designed to allow particles to settle out of suspension. In these circumstances careful base cleaning should be carried out before placing concrete.

A polymer consists of a number of individual molecules joined together to form a chain. The individual molecules are called monomers, and can be joined together to form chains of different lengths and different configurations. Chains can be long or short, and may have branches or be cross-linked to form more complex structures. They may also be coiled. The length and shape of the chains forming the polymer determine its rheological properties and the function it can perform.

The method of specifying the type of polymer required to perform a particular function should be based upon proven performance under simulated and field conditions. There are several brands of polymer available, and they each have different characteristics demanding different control measures. It is essential that the Contractor can demonstrate from trials or from previous experience with the particular brand and its associated additives that the fluid requirements can be achieved.

The polymer should disperse fully in the mixing water before hydration commences, otherwise 'fish eyes' will form. These are lumps of dry material protected from exposure to the mixing water by a layer of the hydrated product. Once formed, the dry material on the inside may fully resist hydration, therefore this material will not be available to develop the fluid properties. To avoid this problem, manufacturers can coat the grains of the polymer with a hydrophobic chemical which, after dispersion, is removed by chemical reaction, shear, temperature or time.

Polymers are used in lower concentrations than bentonite. Typically, the quantity of polymer required to produce a competent support fluid is only about 5–10% of the quantity of bentonite required. However, polymers are significantly more expensive than bentonite and therefore require careful control to minimize losses during use. Polymer slurries require a rigorous testing regime which is more rigorous than for bentonite slurries. In some ground conditions the polymer system appears to be able to revert to a water:soil suspension easier than a bentonite slurry.

Before disposal, the polymer can be broken by the addition of oxidizing agents such as bleach, and the entrained solids allowed to settle. Provided the polymer breaking is done appropriately the remaining fluids are then usually environmentally acceptable to local statutory authorities for disposal into sewers.

C18.3 Evidence of suitability of support fluid

The requirement for submission of details of the support fluid 14 days before commencement of work may prove difficult, especially for fast track projects. Any tests to be carried out on water and bentonite slurries are generally simple and can be carried out shortly before the works start, and still allow time for mix adjustments if necessary.

For a polymer slurry this would be dependent upon the Contractor's experience and the ground conditions.

Support fluids are subject to differing requirements at different stages of construction. During excavation a more viscous and dense fluid will provide greater support to the trench or bore. During concreting, however, a heavier mix may not be fully displaced by the concrete. Under certain conditions the Contractor may elect to use two different mixes of support fluid for the excavation and concreting phases of construction. These would be subject to different compliance values reflecting the differing performance requirements.

C18.4 Materials
C18.4.1 Water

Water for support fluids should come from a potable water supply. If this is not available, alternative sources are likely to require chemical analyses and probably trials to verify their suitability.

Measurement of water by either volume or using a flow meter is a critical element in the production of a consistent well-mixed support fluid. The Contractor should demonstrate either by a calibration certificate or dimensional calculation the method of water measurement prior to commencement of mixing on each project.

C18.4.2 Additives to the water

The requirement for separate waterproof stores may not always be viable for economic or working space reasons. Alternatives, such as shrink wrapped materials on pallets, have been found to perform well, provided appropriate precautions are taken once the shrink wrapping is removed.

Not all materials will comply with Publication No. 163, but may nevertheless be suitable alternatives. Assessment of these materials may be based on evidence of past performance or site trials.

C18.6 Compliance testing of support fluid

Hutchinson *et al.* (1975) and Jefferis (1992) report on the various tests available to measure the properties of bentonite slurry and list appropriate tests. These consist principally of the density, plastic viscosity, shear strength, filtration or fluid loss properties, pH and sand content. Knowledge of these properties is relevant to the following requirements of the fluid:

- solid particles are kept in suspension
- fluid can be easily displaced during concreting
- continuous support of the excavation by fluid pressure acting against a filter cake or a seal created by rheological blocking.

The tests are described in the FPS publication *Bentonite Support Fluids in Civil Engineering in the UK*. Compliance should be checked for the support fluid as supplied to the excavation and prior to concreting because the test requirements of the supporting fluid are different when supporting the excavation compared with the flow conditions needed for effective concreting.

It is impracticable to test each batch of support fluid, which is commonly mixed in small batches to be added to considerably larger stocks of fresh fluid. Tests should, however, be carried out at regular intervals (e.g. daily or more frequently if necessary) on fresh fluid, to ensure the consistency of the batching process.

Fine sand in the slurry during excavation may assist the blocking mechanism. However, the increase in density and viscosity due to the presence of sand may affect the flow properties and hence the ability of the concrete to displace the fluid during concreting. Also, any sediment forming on the base of the excavation may affect the performance of the pile or wall under load if the end bearing

Table C18.1 Typical tests and compliance values for support fluid prepared from bentonite manufactured in the UK

Property to be measured	Test method and apparatus	Compliance values measured at 20°C		
		Freshly mixed	Ready for re-use	Sample from excavation prior concreting
Density	Mud balance	<1.10 g/ml	<1.25 g/ml	<1.15 g/ml
Fluid loss (30 minute test)	Low-temperature test fluid loss	<30 ml	<50 ml	n/a
Filter cake thickness	Low-temperature test fluid loss	<3 mm	<6 mm	n/a
Viscosity	Marsh cone	30–50 seconds	30–60 seconds	30–50 seconds
Shear strength (10 min gel strength)	Fann viscometer	4–40 N/m^2	4–40 N/m^2	4–40 N/m^2
Sand content	Sand screen set	n/a	n/a	<4%*
pH	Electrical pH meter to BS 3445; range pH 7 to 14	7–10.5	7–11	n/a

* 2% prior to concreting if working loads are to be partly resisted by end bearing

component is significant. It is recommended that the sand content should be limited to 2% prior to concreting if working loads are to be partly resisted by end bearing.

Compliance values for a polymer slurry will be quite different to those in Table C18.1.

C19 General requirements for concrete and steel reinforcement

C19.1 General

This Specification takes account of the requirements of BS EN 1536, BS EN 1538 and BS EN 12794 as well as recent changes to British Standards and the introduction of European Standards for concrete materials, specification, production and testing. Other changes have been made to reflect recent changes in concrete technology and the more general use of ready-mixed concrete.

The main factors affecting the successful performance of piling concrete are:

- materials and their availability
- transportation and placing methods
- consistence and compaction
- strength
- durability.

It is most important that all these factors be given equal consideration when considering specification and design.

BS 4449 has been revised to be used in accordance with the European Standard for reinforcement BS EN 10080. This has been necessary as BS EN 10080 does not define steel grades or technical classes. It is recommended that the reinforcing steel should be manufactured and supplied to a recognized third party product certification scheme, such as CARES, meeting the requirements of Clause 8.2 of BS 4449.

BS 4449 covers only steel with characteristic yield strength of 500 MPa.

BS EN 1536 contains requirements for minimum longitudinal reinforcement, size of transverse reinforcement, bar layout, minimum spacing and cover. Particular attention should be paid to the minimum spacing between bars, including other embedded objects such as reservation pipes, to ensure that the free flow of concrete is not impeded.

C19.2.1 Strength class

The minimum strength class of structural concrete should be chosen from Table C19.1.

Table C19.1 Compressive strength classes for structural concrete

Compressive strength class	Minimum characteristic cylinder strength (N/mm^2)	Minimum characteristic cube strength (N/mm^2)
C16/20[1]	16	20
C20/25[1]	20	25
C25/30[1]	25	30
C28/35	28	35
C30/37[2]	30	37
C32/40	32	40
C35/45	35	45
C40/50	40	50
C45/55	45	55
C50/60	50	60

[1] Not suitable for pre-cast piles
[2] This class is not in common use with UK concrete suppliers but may be encountered in European design standards

There is no minimum strength requirement for durability of unreinforced concrete in any exposure condition other than contact with sea water. In this case the minimum strength class is given in Table A.16 of BS 8500-1 and depends on the cement or combination type.

There is no minimum strength class for reinforced concrete in aggressive ground (DC-2 and above) in BRE SD1 or BS 8500-1. Nevertheless, reinforced concrete in non-aggressive ground is subject to a minimum strength class of C25/30 due to the classification of these conditions as XC2. There is no technical basis for this disparity. When in contact with de-icing salt or sea water, the required minimum strength class may be higher.

C19.2.2 Composition of concrete

Additions — normally fly ash or ground granulated blast furnace slag (ggbs) — are in common use in combinations with Portland cement (CEM I) or, less commonly, as factory blended cements for cast-in-situ piling concrete. Their use can infer several advantages and should be used, wherever possible and practical, for concrete in the ground. They enhance properties of the concrete including longer-term strength gain and increased resistance to aggressive ground conditions. They can also be used to impart special properties (e.g. high ggbs content of female piles allows slower early strength development in secant wall construction). They also provide a beneficial material for improvement of self-compacting properties, lower heat of hydration and a more cost-effective material than Portland cement-only concrete.

In addition to minimum cement content requirements, BS EN 1536 requires a minimum fines content (<0.125 mm, including cement and additions) of at least $400 \, \text{kg/m}^3$ for concrete with a nominal maximum aggregate size of greater than 8 mm.

BS 8500 no longer refers to pfa (pulverised fuel ash) and solely uses the term fly ash. Pfa to BS 3892-1 has now been replaced by fly ash to BS EN 450.

Naturally occurring, non-crushed river gravel, e.g. Thames Valley and Trent Valley gravel, are beneficial to the self-compacting performance of cast-in-situ concrete. Crushed aggregates are beneficial to the impact performance of pre-cast concrete piles.

Recycled aggregate is not commonly available as a concrete aggregate but consideration may be given to its use where available. BS 8500-2 classes recycled aggregate as RA or RCA and gives requirements covering its use in concrete. RCA is essentially crushed concrete and is suitable for replacement of approximately 20–40% by mass of the coarse aggregate in concrete of strength class up to C40/50. It is not currently permitted in aggressive ground conditions of DC-2 or above. The use of RA or fine recycled aggregate is not recommended in concrete for piles or embedded retaining walls. Recycled aggregate is, by definition, a crushed material and hence its particle shape may be unsuitable for some uses (e.g. CFA piles).

Concrete should be tested for its susceptibility to bleeding as part of the trial mix procedure in Clause B19.5.2. No guidance is currently available on suitable specification limits for bleeding but experience can easily be gained through routine testing or by special trials of known suitable and unsuitable mixes. Bleed should be minimized by careful mix design to provide a concrete of suitably high consistence but with sufficient cohesion and viscosity to prevent segregation. Propensity to bleed may be increased by the use of excessive water content or insufficient fine material. The specialist use of additions and high range water reducing admixtures (super-plasticizers) may assist in control of bleed.

C19.2.3 Special requirements for non-structural concrete and self-hardening slurry

Plastic concrete or mortar is the term used in BS EN 1538 for 'soft mix' concrete which incorporates clay or bentonite in addition to cement. Typical compositions for plastic concrete and plastic mortar (wherein the aggregate is restricted to fine aggregate) using bentonite are given in Annex A of BS EN 1538. Pre-mixed materials are available for which specification as a Proprietary Concrete in accordance with Clause 4.6 of BS 8500-1 may be appropriate.

BS 8500 only lists cement combinations containing up to 55% fly ash, or up to 80% ggbs by mass of the combination. Nevertheless, higher proportions may be required in non-structural concrete where low early strength is required.

Information on the performance in aggressive ground conditions of concrete made with cement containing a higher proportion of ggbs than covered by BRE SD1 (i.e. greater than 80% by mass) is given in a BRE Information Paper 2005/17.

C19.2.4 Consistence

The consistence of concrete is critical to the successful construction of cast-in-situ concrete piling. The concrete must be designed as self-compacting since vibration is inappropriate and the concrete may need to flow between closely spaced reinforcement. The use of excessive water to provide high consistence may lead to segregation and excessive bleed and should be avoided. Careful mix design is necessary to ensure minimization of bleed (see Clause B19.2.2), especially for large diameter piles and diaphragm wall panels. So called 'self-compacting concrete' is available from many concrete suppliers for use in conventional structural concrete. These mixes may differ from suitable self-compacting piling mixes as they may have very high cement contents, often comprising plain CEM I Portland cement, and expensive 'new generation' super-plasticizers and viscosity modifying agents.

The measurement of concrete consistence should only be used as an indicator of the suitability of the concrete for placing and compaction, and not as a possible strength indicator.

BS EN 12350-2 states that the slump test may not be suitable for determining consistence where the measured slump exceeds 200 mm. For very high consistence concrete, such as defined by consistence C in Table 3.1, the use of the flow table (to BS EN 12350-5) is recommended. The method of consistence measurement can be specified in the Project Specification.

BS EN 206-1 gives consistence classes for slump and flow but permits specification by target value in special cases. The classes are likely to be too wide for most piling applications and specification by target value may be preferable. The tolerances on target values are given in Table 11 of BS EN 206-1. In addition, a deviation of +20/−10 mm (+30/−20 mm for initial discharge) is allowed on top of the ±30 mm tolerance permitted for any specified target slump greater than 100 mm. This gives a total tolerance of +50/−40 mm (+60/−50 mm for initial discharge) and would be unacceptable for many applications including diaphragm walling.

BS EN 1538 recommends a flow value of between 550 mm and 600 mm for concrete in diaphragm walls. Recommended consistence values from BS EN 1536 for bored piles in various conditions are given in Table C19.2.

Table C19.2 Recommended consistence ranges from BS EN 1536 for bored piles

Flow range (mm)	Slump range (mm)	Examples of applications
460–530	130–180	Placement in dry conditions
530–600	≥160	Placement by pumping Placement by tremie under water
570–630	≥180	Placement by tremie in submerged condition under support fluid

Discussion and agreement of the particular slump or flow requirements, including tolerances, with the concrete supplier is essential before commencing the works.

C19.2.5 Alkali–silica reaction

BS 8500 requires the concrete producer, whether for ready-mixed or site-batched concrete, to minimize the risk of damaging alkali–silica reaction for designed concrete. No further requirement is needed in the Specification.

For prescribed concrete the responsibility remains with the specifier, who is defined as the person or body establishing the final compilation of technical requirements and thus will often be the Contractor. Where prescribed concrete is to be used the requirements and responsibility for minimizing the risk of damaging alkali–silica reaction should be clearly defined in the Project Specification.

C19.3 Ready-mixed concrete

Most concrete for piling applications in the UK is likely to be obtained from ready-mixed concrete suppliers and the Specification has been changed to reflect this. Only those who hold third party accreditation such as membership of the Quality Scheme for Ready Mixed Concrete (QSRMC) are likely to be able to meet the requirements of the Specification.

Under some circumstances (e.g. using high slump mixes in hot weather) the planned and supervised addition of water at site, to control mix properties, is preferable to other measures taken to meet workability limits and should therefore not be ruled out.

C19.4 Site-batched concrete

Under some circumstances (e.g. where very large quantities are required) it may be practical for concrete to be supplied from an on-site batching plant rather than local ready-mixed concrete suppliers. The quality requirements in BS 8500 apply equally to site-mixed concrete as it does to ready-mixed concrete. Where possible and practical, site batching plants should operate under a third-party accredited quality scheme in the same way as ready-mixed concrete. Third-party accreditation will not always be possible but the plant should still operate to the same level of production control as required for ready-mixed concrete plants.

C19.4.2 Materials

Restrictions on materials should be kept to a minimum, and locally available materials should always be specified or permitted for use unless particularly unsuitable.

C19.4.2.1 Cement and additions

The term 'cement' is used in BS 8500 and the Specification to refer to factory produced cement including blended cements and, for those that comply with BS EN 197, are designated by the suffix CEM (e.g. CEM IV/B-V). The term 'combination' refers to the common UK practice of adding Portland cement (CEM I) and an addition, such as pfa or ggbs, separately to the mixer and are normally designated by the suffix 'C' (e.g. C IV/B-V). The suffix 'B' is used for combinations of Portland cement and ggbs which conform to the composition requirements for type IIIA or IIIB cement to BS EN 197-1 but not to the requirements for early strength gain; these cements are covered by BS EN 197-4. BS 8500 recognizes the properties of combinations to be equivalent to factory blended cements of the same composition.

Sulfate-resisting Portland cement (SRPC) is no longer commonly available from ready-mixed concrete suppliers in the UK. Equivalent performance may be obtained by the use of suitable blended cements or combinations with pfa or ggbs in accordance with BRE SD 1.

BS EN 12620 for aggregates for concrete is less restrictive than the previous standard BS 882. A British Standard Published Document, PD 6682-1, contains supplementary requirements for aggregates to achieve equivalence with BS 882 and aggregates for concrete should thus meet the requirements of both BS EN 12620 and PD 6682-1.

C19.4.2.2 Aggregate

Fine aggregates which consist mainly of angular particles are to be avoided. The limits for shell content in Table B19.2 are changed from the previous version of the Specification but are in accordance with the recommendations of PD 6682-1 (see Clause C19.1.1). Maximum shell content should be specified using the BS EN 12620 category.

C19.4.2.3 Water

Potable water is accepted by BS EN 1008 as suitable for use in concrete without the need for testing.

C19.4.2.4 Admixtures

The expert use of plasticizing and superplasticizing admixtures in concrete is encouraged to avoid the use of high free water contents to obtain the required high levels of consistence and to help reduce bleed. The use of retarding admixtures may be necessary to prolong workability as required. Where two or more admixtures are used in combination, evidence should be obtained of their compatibility.

C19.4.3 Batching concrete
C19.4.3.2 Accuracy of weighing and measuring equipment

BS 1305 is current but obsolescent. It has been retained in the Specification because there is no alternative British Standard or European Standard giving accuracy limits for batching equipment.

C19.4.4 Mixing concrete
C19.4.4.1 Type of mixer

BS 3963 is current but obsolescent. It has been retained in the Specification because there is no alternative British or European Standard giving requirements for concrete mixers. Testing is not required unless it is visually apparent that the concrete is not being properly mixed.

C19.4.4.4 Minimum temperature

Flaked ice may be added to the concrete mix to reduce its temperature but mixing time should be increased to ensure it is fully melted by the time of completion of mixing. Crushed ice should not be used unless a screen is used to ensure no larger lumps are added which would not be fully melted during mixing.

C19.5 Trial mixes
C19.5.1 General

It may be the case that there is insufficient time available to carry out trial mixes and under these circumstances the data provided by the concrete producer of previous production of concrete of similar proportions should be taken into consideration. Trial mixes should not then be necessary.

Pre-cast concrete piles are normally produced by an industrially-based rather than project-based process. The pre-cast pile producer's quality scheme should provide current and historical records showing suitability of concrete, and under these circumstances trial mixes are not required.

Where additions are used, consideration should be given to the strength compliance requirements of concrete due to the slower rate but increased duration of strength gain compared to plain Portland cement concrete. Under such circumstances it may well be appropriate to consider a reduction factor on the 28-day strength class, rather than increasing the basic compliance time, e.g. to 56 days (since such a time-scale must be considered impractical for most construction operations).

The number of cubes required from trial mixes is less than that previously required but is in line with BS EN 1536 for bored piles.

C19.5.4 Consistence

The concrete produced should be cohesive and in a slump test should not shear, fragment or segregate easily. Guidance on consistence is also given in Clause C19.1.4.

C19.8 Testing concrete
C19.8.1 Sampling

The point of placing of the concrete should be considered as the point of discharge from the delivery vehicle or mixer, e.g. the point of discharge of concrete into a pump for CFA piling, or a skip or dumper for cast-in-situ piling.

C19.8.2 Consistence testing

See Clause C19.1.4.

C19.8.3 Sampling for compressive strength testing

BS 8500 puts the responsibility for demonstrating conformity of compressive strength with the concrete producer, with the option of 'identity testing' of site cubes by the Contractor. Nevertheless, the rate of sampling required of the concrete producer is insufficient for piling applications so the tradition of site cubes is continued in the Specification. Sampling, specimen preparation, storage and testing may be performed by the concrete producer, the Contractor, an independent third party or any combination of these provided the requirements of the Specification are met.

The specified sampling regime has been changed but is in line with the requirements of BS EN 1536 for bored piles. Should a more onerous rate of sampling be required, this should be specified in the Project Specification. The importance of good sampling, cube making and curing is stressed to avoid any possible false non-conformity of strength due to unrepresentative cubes.

C19.8.4 Compressive strength testing

The specified testing regime has been changed but is in line with the requirements of BS EN 1536 for bored piles. One cube is held in reserve for testing at a later age should the results of the 28 day tests indicate non-conformity. This may be useful in the case of concrete made with blended cement or combinations with fly ash or ggbs that generally undergo significant post 28 day strength gain.

C19.8.5 Acceptance criteria for compressive strength

Assessment of conformity through identity testing is ill-defined in BS 8500 and BS EN 206-1 so to meet the requirements of piling applications the Specification gives conformity requirements based on those in the former British Standard for concrete, BS 5328-4, which has long been accepted and used by the industry. Compressive strength conformity is thus judged on the basis of the running-mean-of-four results. It may be necessary for the Contractor to enter into special arrangements with the concrete producer who will be assessing conformity according to the requirements of BS 8500. Disputes are otherwise foreseeable should site cubes fail to conform but when the concrete producer claims conformity in accordance with the limited tests required by BS 8500.

The acceptance criteria in Table B19.3 of the Specification are more onerous than those in Table B.1 of BS EN 206-1 but are identical to those in the former BS 5328-4.

Where non-conformity of cube strength occurs, due recognition should be given to the factors of particular importance to pile durability rather than just strength, i.e. w/c ratio, adequate placing, adequate compaction, as well as the stresses to which the piles will be subjected in service. Strength should not be considered the arbiter of durability although low strength may be indicative of a higher water/cement ratio than required for durability.

C19.9 Steel reinforcement
C19.9.3 Fabrication of reinforcement

Hoops, links or helical reinforcement for piles may be used in conjunction with proprietary circular steel or plastic cages formers to provide the specified cover to the longitudinal reinforcement.

C19.10 Grout
C19.10.1 General

Clause 5.2 of BS 8500-2: 2006, resistance to alkali–silica reactions, and BRE Digest 330, to which the standard refers for detailed guidance, are intended for use with conventional concrete. Nevertheless, they may be equally applicable to grout containing fine aggregate and with total cement content not greater than $550\,kg/m^3$. No specific guidance is available for grout containing greater than $550\,kg/m^3$ but careful selection of constituents should allow a satisfactory material to be produced, particularly with low reactivity aggregates.

Grout containing no aggregate is not at risk from alkali–silica reaction because the source of reactive silica is invariably the aggregate.

The principle in BRE Digest 330 for minimization of the risk of damaging alkali–silica reactions is to limit the total alkali content of the mix to a level dependent on the reactivity classification of the aggregate. Aggregate reactivity is classed as low, medium or high by consideration of the mineralogical composition or, where necessary, by testing. Testing is unlikely to be necessary for aggregates already in common use.

The detailed procedures in BRE Digest 330 for ensuring the total alkali content is within acceptable limits depend on the certified alkali content of the cement, the use of any cementitious additions such as fly ash or ggbs, the reactivity classification of the aggregate, and the likelihood of additional alkalis from other sources. The certified alkali content of the cement and the reactivity classification of the fine aggregate should be available from the suppliers. The use of fly ash and ggbs, in sufficient proportion (generally at least 25% by mass of cement for fly ash or at least 40% by mass of cement for ggbs, but higher proportions for use with highly reactive aggregates), can be very effective in mitigating the risk of damaging alkali–silica reaction.

It should not normally be necessary to perform routine testing for alkali–silica reaction during the course of a contract.

C19.10.2 Batching

BS 1305 is still current but classed by BSI as obsolescent. It has been retained in the specification because there is no alternative British or European Standard giving accuracy limits for batching equipment.

C19.11 Form for specification of designed concrete

1. Concrete designation.
2. Strength class — this should be appropriate to the structural requirements except where durability requirements dictate a higher class.
3. DC class — where the DC class is specified it is not be necessary to specify requirements for maximum water/cement ratio or minimum cement content, if they are to be the same as given by BRE SD1, as these are automatically called up by the requirement to comply with BS 8500.
4. Maximum water/cement ratio — see item 3 above.
5. Minimum cement content — see item 3 above and Clause B19.1.2 of the Specification.
6. Permitted cement and combination types.
7. Nominal maximum size of aggregate (mm) — the concrete supplier will normally assume 20 mm unless specified otherwise. If 10 mm is specified, the minimum cement content may have to be adjusted in accordance with Table A of BS 8500-1.
8. Chloride class — see B19.1.2 and Table B19.1 of the Specification.
9. Special requirements for cement or combination — it may be desirable to specify limitations on addition content within those given for particular cement or combination types.
10. Special requirements for aggregates — rounded aggregates may be desirable for some applications.
11. Special requirements for temperature of fresh concrete — only necessary if required to be greater than 5°C or less than 35°C.
12. Special requirements for strength development — see Clause 7 of BS 8500.
13. Special requirements for heat development during hydration.
14. Other special technical requirements — e.g. retardation of stiffening.
15. Additional requirements.
16. Rate of sampling for strength testing — see Clause B19.5.3.
17. Whether the use of recycled aggregate (RCA) is permitted.
18. Target consistence or consistence class — see C19.1.4.
19. Tolerance on target consistence if different from BS EN 206-1 — see Clause C19.1.4.
20. Method of placing concrete — this is to be entered by the Contractor for the information of the concrete supplier.

References

American Petroleum Institute Specification for Line Pipes 5L. 43rd edition. API. 2004.

Arcelor RPS. *Piling Handbook 8th ed.* Arcelor. 2005.

Arcelor RPS. *The impervious steel sheet pile wall Parts 1 and 2.* Arcelor. 2006.

Arcelor RPS. *Welding of Steel Sheet Piles.* Arcelor. 2006.

ASTM D 4945-00. *Standard test method for high-strain dynamic testing for piles.*

British Standards Institution. British Standard 144: *Specification for coal tar creosote for wood preservation.* BSI, London. 1997.

British Standards Institution. British Standard 882: *Specification for aggregates from natural sources for concrete.* BSI, London. 1983.

British Standards Institution. British Standard 450-1. *Fly ash for concrete. Definition, specification and conformity criteria.* BSI, London. 2005.

British Standards Institution. British Standard 1047: *Specification for air-cooled blast furnace slag aggregate for use in construction.* BSI, London. 1983.

British Standards Institution. British Standard 1282: *Wood preservatives. Guidance on choice, use and application.* BSI, London. 1999.

British Standards Institution. British Standard 1305: *Specification for batch type concrete mixers.* BSI, London. 1974.

British Standards Institution. British Standard 3892-1: Pulverised-fuel ash. Specification for pulverised-fuel ash for use with Portland cement. BSI, London. 1997.

British Standards Institution. British Standard 3900-E6 *Paints and varnishes. Cross-cut test.* BSI, London. 1995.

British Standards Institution. British Standard 3963: Method testing the mix performance of concrete mixers. BSI, London. 1974.

British Standards Institution. British Standard 4072: Copper/chromium/arsenic preparations for wood preservation. BSI, London. 1999.

British Standards Institution. British Standard 4449: *Steel for the reinforcement of concrete. Weldable reinforcing steel. Bar, coil and decoiled product.* Specification. BSI, London. 2005.

British Standards Institution. British Standard 4978: *Specification for visual strength grading of softwood.* BSI, London. 1996.

British Standards Institution. British Standard 5228: Noise control on construction and open sites, Part 4: code of practice for noise and vibration control applicable to piling operations. BSI, London. 1992.

British Standards Institution. British Standard 5328-4: *Concrete.* BSI, London. 1997.

British Standards Institution. British Standard 5756: *Specification for visual strength grading of hardwood.* BSI, London. 1997.

British Standards Institution. British Standard 5930: *Code of practice for site investigations.* BSI, London. 1999.

British Standards Institution. British Standard 6682-1: *Method for determination of bimetallic corrosion in outdoor exposure corrosion tests.* BSI, London. 1989.

British Standards Institution. British Standard 7079-A1: *Preparation of steel substrates before application of paints and related products. Visual assessment of surface cleanliness. Rust grades and preparation grades of uncoated steel substrates and of steel substrates after overall removal of previous coatings.* BSI, London. 1989.

British Standards Institution. British Standard 7385: *Evaluation and measurement for vibration in buildings.* BSI, London. 1990.

British Standards Institution. British Standard 7973-1: *Spaces and chairs for steel reinforcement and their specification, product performance requirements.* BSI, London. 2001.

British Standards Institution. British Standard 7973-2 *Spaces and chairs for steel reinforcement and their specification. Fixing and application of spaces and chairs and tying of reinforcement.* BSI, London. 2001.

British Standards Institution. British Standard 8102: *Code of practice for protection of structures against water from the ground,* BSI, London, 1990.

British Standards Institution. British Standard 8110: *Structural use of concrete.* BSI, London. 1997.

British Standards Institution. British Standard 8417: *Preservation of timber, recommendations.* BSI, London. 2003.

British Standards Institution. British Standard 8500: *Concrete. Complementary British Standard to BS EN 206-1.* BSI, London. 2006.

British Standards Institution. British Standard 8500-1: *Concrete. Complementary British Standard to BS EN 206-1. Method of specifying and guidance for the specifier.* BSI, London. 2006.

British Standards Institution. British Standard 8500-2: *Concrete. Complementary British Standard to BS EN 206-1. Specification for constituent materials and concrete.* BSI, London. 2006.

British Standards Institution. British Standard 8666: *Scheduling dimensioning, bending of steel reinforcements for concrete specification.* BSI, London. 2005.

British Standards Institution. British Standard BS 4-1. *Structural steel sections. Specification for hot-rolled sections.* BSI, London. 2005.

British Standards Institution. British Standard BS 450-1: *Fly ash for concrete. Definitions, specification and conformity criteria.* BSI, London. 2005.

British Standards Institution. British Standard BS 8102: *Code of practice for protection of structures against water from the ground.* BSI, London. 1990.

British Standards Institution. British Standard EN 197-1: *Cement. Composition, specification and conformity criteria for common cements.* BSI, London. 2000.

British Standards Institution. British Standard EN 197-4: *Cement. Composition specifications and conformity criteria for low early strength blastfurnace cements.* BSI, London. 1998.

British Standards Institution. British Standard BS EN 206-1: *Concrete. Specification, performance, production and conformity.* BSI, London. 2000.

British Standards Institution. British Standard EN 287-1: *Qualification test of welders. Fusion welding. Steels.* BSI, London. 2004.

British Standards Institution. British Standard EN 287: *Approval testing of welders for fusion welding.* Parts 1 and 2. 1992.

British Standards Institution. British Standard EN 350-1: *Durability of wood and wood based products. Natural durability of solid wood. Guide to the principles of testing and classification of natural durability of wood.* BSI, London. 1994.

British Standards Institution. British Standard EN 350-2: *Durability of wood and wood-based products. Natural durability of solid wood. Guide to natural durability and treatability of selected wood species of importance in Europe.* BSI, London. 1994.

British Standards Institution. British Standard EN 480-4: *Admixtures for concrete, mortar and grout. Test methods. Determination of bleeding in concrete.* BSI, London. 2005.

British Standards Institution. British Standard EN 934-2: *Admixture for concrete, mortar and grout. Concrete admixture. Definitions, requirements, conformity, marking and labelling.* BSI, London. 2001.

British Standards Institution. British Standard EN 1008: *Mixing water for concrete. Specification for sampling, testing and assessing the suitability of water, including water recovered from processes in the concrete industry, as mixing water for concrete.* BSI, London. 2002.

British Standards Institution. British Standard EN 1011-1: *Welding recommendations for welding of metallic materials, general guidance for arc welding.* BSI, London. 1998.

British Standards Institution. British Standard EN 1011-2: *Welding recommendations for welding of metallic materials. Arc welding of ferretic steels.* BSI, London. 2001.

British Standards Institution. British Standard EN 1536: *Execution of special geotechnical work piles.* BSI, London. 2000.

British Standards Institution. British Standard EN 1561: *Founding, grey cast irons.* BSI, London. 1997.

British Standards Institution. British Standard EN 1563: *Specification, cast iron sectional tanks (rectangular). Founding. Spheroidal graphite cast iron.* BSI, London. 1997.

British Standards Institution. British Standard EN 1583: *Execution of special geotechnical work – diaphragm walls.* BSI, London. 2000.

British Standards Institution. British Standard EN 1990: *Eurocode basis of structural design.* BSI, London. 2002.

British Standards Institution. British Standard EN 1991-1-2: Eurocode 1: *Actions on structures,* BSI, London. 2002.

British Standards Institution. British Standard EN 1992-1-1: Eurocode 2: *Design of concrete structures. General rules and the rules for buildings.* BSI, London. 2004.

British Standards Institution. British Standard EN 1992-3. Eurocode 2: *Design of concrete structures: concrete foundations.* BSI, London. 1998.

British Standards Institution. British Standard EN 1993-5. Eurocode 3: *Design of steel structures, Part 5: Piling.* BSI, London. 2006.

British Standards Institution. British Standard EN 1995-1-1. Eurocode 5: *Design of timber structures, General, common rules and rules for buildings.* BSI, London. 2004.

British Standards Institution. British Standard EN 1997-1: 2004 Eurocode 7: *Geotechnical design, Part 1: General rules.* BSI, London. 2004.

British Standards Institution. British Standard EN 1997-2: 2007 Eurocode 7: *Geotechnical design, Part 2: Ground investigation and testing.* BSI, London. 2007.

British Standards Institution. British Standard EN 5817: *Welding. Fusion-welded joints in steel, nickel, titanium and their alloys (beam welding excluded). Quality levels for imperfection.* BSI, London. 2003.

British Standards Institution. British Standard EN 10024: *Hot rolled taper flange I sections. Tolerances on shape and dimensions.* BSI, London. 1995.

British Standards Institution. British Standard EN 10025-1: *Hot rolled products of structural steels. General technical delivery conditions.* BSI, London. 2004.

British Standards Institution. British Standard EN 10034: *Structural steel I and H sections. Tolerances on shape and dimensions.* BSI, London. 1993.

British Standards Institution. British Standard EN 10210-1: *Hot finished structural hollow sections of non-alloy and fine grain steels. Technical delivery requirements.* BSI, London. 2006.

British Standards Institution. British Standard EN 10210-2. *Hot finished structural hollow sections of non-alloy and fine grain steels. Tolerences, dimensions and sectional properties.* BSI, London. 2006.

British Standards Institution. British Standard EN 10219-1: *Cold formed welded structural hollow sections of non-alloy and fine grain steels. Technical delivery requirements.* London. 2006.

British Standards Institution. British Standard EN 10219-2. *Cold formed welded structural hollow sections of non-alloy and fine grain steels. Tolerences, dimensions and sectional properties.* BSI, London. 2006.

British Standards Institution. British Standard EN 10248-1. *Hot rolled sheet piling of non alloy steels. Technical delivery conditions.* BSI, London. 1996.

British Standards Institution. British Standard EN 10248-2. *Hot rolled sheet piling of non alloy steels. Tolerances on shape and dimensions.* BSI, London. 1996.

British Standards Institution. British Standard EN 10249-1: *Cold formed sheet piling of non alloy steels. Technical delivery conditions.* BSI, London. 1996.

British Standards Institution. British Standard EN 10249-2: *Cold formed sheet piling of non alloy steels. Tolerances on shape and dimensions.* BSI, London. 1996.

British Standards Institution. British Standard BS EN 10293: *Steel castings for general engineering uses.* BSI, London. 2005.

British Standards Institution. British Standard EN 12063: *Execution of special geotechnical work. Sheet pile walls.* BSI, London. 1999.

British Standards Institution. British Standard EN 12350-1: *Testing fresh concrete. Sampling.* BSI, London. 2000.

British Standards Institution. British Standard EN 12350-2: *Testing fresh concrete. Slump test.* BSI, London, 2000.

British Standards Institution. British Standard EN 12350-5: *Testing fresh concrete. Flow table test.* BSI, London. 2000.

British Standards Institution. British Standard EN 12390-1: *Testing of hardened concrete.* BSI, London. 2000.

British Standards Institution. British Standard EN 12620: *Aggregates for concrete.* BSI, London. 2002.

British Standards Institution. British Standard EN 12699: *Execution of special geotechnical work, displacement piles.* BSI, London, 2001.

British Standards Institution. British Standard EN 12794: *Precast concrete products, foundations piles.* BSI, London. 2005.

British Standards Institution. British Standard EN 14081-1: *Timber structures. Strength graded structural timber with rectangular cross section. General requirements.* BSI, London. 2005.

British Standards Institution. British Standard EN 14199: *Execution of special geotechnical works. Micropiles.* BSI, London. 2005.

British Standards Institution. British Standard EN 15614-1: *Specification and qualification of welding procedures for metallic materials. Welding procedure test. Arc and gas welding of steels and arc welding of nickel and nickel alloys.* BSI, London. 2004.

British Standards Institution. British Standard EN ISO 2409: *Paints and varnishes. Cross-cut test.* BSI, London. 1995.

British Standards Institution. British Standard EN ISO 8501-1: *Preparation of steel substrates before application of paints and related products. Visual assessment of surface cleanliness. Rust grades and preparation grades of uncoated steel substrates and of steel substrates after overall removal of previous coatings.* BSI, London. 2001.

British Standards Institution. British Standard EN ISO 9001: *Quality management systems, requirements.* BSI, London. 2000.

British Standards Institution. British Standard EN ISO 12944: *Paints and varnishes.* BSI, London. 2000.

British Standards Institution. British Standard EN ISO 15607: *Specification and qualification of welding procedures for metallic materials. General rules.* BSI, London. 2003.

British Standards Institution. British Standard EN ISO 15609-1: *Specification and qualification of welding procedures for metallic materials building procedures. Specification arc welding.* BSI, London. 2004.

British Standards Institution. British Standard EN ISO 15613: *Specification and qualification of welding procedures for metallic materials. Qualification based on pre-production welding test.* BSI, London. 2004.

British Standards Institution. British Standard EN ISO 22477-1. *Pile load test by static axially loaded compression.* BSI, London. (In production)

British Standards Institution. Published Document 6682-1 Aggregates. *Aggregates for concrete. Guidance on the use of BS EN12620.* BSI, London. 2003.

Building Research Establishment. *Working for Tracked Plant.* BR 470. BRE. 2004.

Building Research Establishment. BRE Digest 330: *Alkali–Silica reaction in concrete.* BRE. 2004.

Building Research Establishment. BRE Digest 479: *Timber pile and foundations.* BRE. 2003.

Building Research Establishment. BRE Special Digest 1: *Concrete in aggressive ground.* BRE. 2005.

Building Research Establishment. Information Paper 2005/17: *Concrete with high ggbs contents for use in hard/firm secant piling.* BRE. 2005.

Building Research Establishment. Report BR470: *Working platforms for tracked plant.* BRE. 2004.

Chin, F. K., (1970) *Estimation of the ultimate load of piles from tests not carried to failure.* Proc. 2nd S E Asian Conf. Soil Engineering, Singapore, 81–92.

Construction (Design and Management) Regulations (CDM). Statutory Instrument 1994 No. 3140. Crown Copyright. 1994. (Amended 2000)

Construction (Design and Management) Regulations (CDM). Statutory Instrument 2007 No. 320. Crown Copyright. 2007.

Construction Industry Research and Information Association. *CDM Regulations – work sector guidance for designers.* CIRIA, Report 166. 1997.

Construction Industry Research and Information Association. *Integrity testing in piling practice.* CIRIA, Report R144. 1997.

Construction Industry Research and Information Association. *Water-resisting basements.* CIRIA, Report 139. 1995.

Construction Industry Research and Information Association. CIRIA, Report PG3. The use and influence of bentonite in bored pile construction. CIRIA. 1997.

Construction Industry Research and Information Association. CIRIA, Report PG7. *Pile load testing procedure.* CIRIA. 1980.

Construction Industry Research and Information Association. DOE and CIRIA piling development group report PG2: Review of problems associated with the construction of cast-in-place concrete piles. CIRIA. 1977.

Construction Industry Research and Information Association. Gaba, A. R., Simpson, B. Powrie, W. and Beadman, D. R. *Embedded retaining walls – guidance for economic design.* CIRIA, Report C580. 2003.

Creosote Regulations 2003. Statutory Instrument 2003 No. 1511. The Creosote (Prohibition on Use and Marketing) (No. 2) Regulations. www-opsi.gov.uk

Environment Protection Regulations. 2003. Statutory Instrument 2003 No. 63. The Environment Protection (Duty of Care) (England) (Amendment) Regulations. www-opsi.gov.uk

Federation of Piling Specialists. *Bentonite support fluids in civil engineering in the UK.* 2nd ed. www.fps.org.uk

Federation of Piling Specialists. *Essential guide to the ICE specification for piling and embedded retaining walls.* Thomas Telford, London. 1999.

Federation of Piling Specialists. *Handbook on pile load testing.* British Standard. www.fps.org.uk 2006.

Fleming, M. and England W. G. K. (1994). Review of foundation testing methods and procedures. *Proc. ICE Geotechnical Engng*, paper 10407.

Fleming, W. G. K. (1992) *A New method for single pile settlement prediction and analysis.* Geotechnique, XLII, 3 411–425.

Highways Agency. *Specification for piling and embedded retaining walls.* HMSO. 1994.

Hutchinson, M. T, Daw, G. P. Shotton, P. G. and James, A. N. The properties of bentonite slurries used in diaphragm wall and their control. *Proc. Conf. on diaphragm walls and anchorages.* ICE, London. 1975, 33–40.

Institution of Civil Engineers. *Specification for piling and embedded retaining walls.* Thomas Telford, London. 1996.

Institution of Civil Engineers. *Specification for piling*. Thomas Telford, London. 2004.

Jefferis, S. (1992) Grout and Slurries. The Construction Materials Reference Book. Doran, D. K., Ed. Butterworth.

Leung, D. F., Tan, S. A. and Phoon, K. K. (1999) Proceedings of the International symposium on field measurements in geomechanics, 5, Singapore, 1–3 Dec. 1999.

Myrovoll, F. (2003) Proceedings of the International symposium on field measurements in geomechanics, 6, Oslo, 15–18 Sept. 2003.

North American Steel Sheet Piling Association. *Sheet Piling TESPA Installation Guide*. NASSPA. 2005.

Proceedings of the International symposium on field measurements in geomechanics, 4, Bergamo, 10–12 Apr. 1995.

Publication no 163 (1988). *Drilling Fluid Materials*. The Engineering Equipment and Material Users Association. www.eemua.org

Puller, M. J. (1994) The Waterproofness of Structural Diaphragm Wall. *Proc. Inst. Civ. Eng*, 107, pp. 47–57.

Sörum, G. (1991) *Proceedings of the International symposium on field measurements in geomechanics*, 3, Oslo, 9–11 Sept. 1991 vol. 1–2.

The marketing and use of dangerous substance regulations 2003 SR 2003/105. www-opsi.gov.uk